Chemical Inhibitors for Corrosion Control

Special Publication No. 71

Chemical Inhibitors for Corrosion Control

The Proceedings of an International Symposium, Organized by the Industrial Division of the Royal Society of Chemistry and the Institution of Corrosion Science and Technology

University of Manchester, 21st–22nd April 1988

Edited by
B. G. Clubley
Ciba-Geigy Industrial Chemicals
Manchester

ROYAL
SOCIETY OF
CHEMISTRY

British Library Cataloguing in Publication Data

Chemical inhibitors for corrosion control: the proceedings of an international symposium organised by the Industrial Division of the Royal Society of Chemistry and the Institution of Corrosion Science and Technology, University of Manchester, 21–22 April 1988.
1. Metals. Corrosion inhibitors
I. Clubley, B. G. II. Royal Society of Chemistry, *Industrial Division* III. Institution of Corrosion Science and Technology IV. Series
620.1623

ISBN 0–85186–836–3

Published by the Royal Society of Chemistry, Thomas Graham House, Science Park, Cambridge CB4 4WF

Printed in Great Britain by Henry Ling Ltd., at the Dorset Press, Dorchester, Dorset

Preface

Corrosion is one of the major problems affecting the performance, safety and appearance of materials. Despite continuing advances in corrosion resistant materials, the use of chemical inhibitors often remains the most practical and cost effective means of preventing corrosion.

The use of chemicals to inhibit corrosion of metals in many environments is well established. However, although effective products are in use for many applications, this practice is in a state of change.

There are, in some areas, demands for improved performance of inhibitors to operate under more severe conditions. A greater driving force comes from increasing concern for the environment and more awareness of the adverse effect of certain chemicals. Today's inhibitors, in common with other chemical products, must maintain or improve on existing technical performance. They must be safe to use and environmentally acceptable. They must also have the versatility to cope with new applications and formulations.

Improved products can only be developed efficiently from greater understanding of corrosion and inhibition mechanisms. Particularly relevant are studies under 'real life' conditions, the effect of flow and the understanding of localised corrosion. Further use of modern investigative techniques is required to give information on mechanism, performance and inhibitor monitoring.

The subject of 'Chemical Inhibitors for Corrosion Control' was addressed at a two day international conference held at the University of Manchester Institute of Science and Technology on 21st and 22nd April 1988. This meeting was organised jointly by the Royal Society of Chemistry Industrial Division and the Institution of Corrosion Science and Technology.

The meeting was attended by 130 delegates from the U.K. and abroad. Speakers from industry and universities presented papers on different types of inhibitors and their use, testing and mechanism. A wide range of applications including oil fields, cooling water systems, steam generating systems, coatings and lubrication services were covered.

This monograph describes the proceedings of the meeting including papers and poster session.

I wish to acknowledge the work of the other members of the organising committee namely, Dr. J. Dawson (U.M.I.S.T.), Dr. D. Harrop (B.P.), Dr. P. Ogden (R.S.C.), Dr. N. Rothwell (CAPCIS), Dr. J. Wolstenholme (Newtech) and particularly Dr. S. Gibson (Akzo Chemie) whose efficient organisation ensured the success of the meeting.

I also wish to thank the authors for their presentations and for providing their manuscripts, U.M.I.S.T. for providing the conference facilities and Mr. A.G. Cubitt and Mrs. R.H. Pape of the R.S.C. for arranging the publication.

I hope this monograph will prove a useful guide to the current position of chemical inhibitors for corrosion control both for those with experience in the field and for newcomers wishing to learn.

B.G. CLUBLEY

Contents

Chemical Inhibitors for Corrosion Control

By D. Harrop

BP RESEARCH CENTRE, SUNBURY-ON-THAMES, MIDDLESEX TW16 7LN, UK

1 INTRODUCTION

In many industries, the need to use constructional materials safely, but cost-effectively, is a primary consideration. Frequently, physical requirements can be satisfied easily, but corrosion effects seriously complicate the selection of suitable materials.

Generally, increased corrosion-resistance can only be obtained at increased cost. However, the actual material-related costs incurred in a project will depend on the corrosivity of the environment concerned, the required design life, the physical requirements of the material and the ease with which it can be worked or machined, and readily available stocks. In some cases, appearance may also dictate the use of a more expensive material.

At a fundamental level, corrosion-resistant materials quite often tend to be more susceptible to localised corrosion, even though they may withstand general attack. Localised corrosion can be a far more insidious form of deterioration, and, under certain circumstances, can promote catastrophic failure (e.g. stress corrosion cracking). Localised corrosion is also often difficult to detect early and to monitor effectively.

The costs and problems associated with corrosion-resistant materials means that, in many cases, the use of corrosion inhibitors is a practical and economic alternative, which can be designed-in, or a retrofitted treatment. Industrial use of corrosion inhibitors is, therefore, now broad based and extensive. A recent survey[1] estimated that the US industry spent $798 M (1.54 M tonnes) on corrosion inhibitor chemicals in 1986. The figure for the EEC for 1986 was probably ca.$150 M.

This paper attempts to present an overview of chemical corrosion inhibitors as a broader introduction for the more specific symposium papers which follow, introducing principles and nomenclature with some general comments on practice.

2 CORROSION INHIBITOR MARKET

The US and EEC markets for corrosion inhibitors are significantly different in their make up. Figures 1 and 2 summarise analysis of these markets[2] which, apart from the difference in overall size, the most striking difference lies in how the market share is split: oil production and refining taking the major share of the US market, whereas the EEC market is dominated by water treatment. Overall percentage annual growth of the two markets is similar, with oil production/drilling/refining predicted to be the biggest growth sector for both the US and EEC markets. Other parallels are difficult to draw, but the validity of the predictions may be viewed against the most recent survey of the US market estimated at $798M for 1986[1]. This agrees remarkably well with that predicted based on the 1981 figures and perceived market trends (Figure 1). However, what effect the major fall in the price of oil in 1986 will have on forward growth figures remains to be seen.

The 1986 US survey highlights lime, chromates and phosphates within the 39% ($311M) inorganic chemicals share of the market; amidoamines, imidazolines, fatty acids, petroleum sulphonates, phosphonates, alkanolamines and neutralising amines feature in the 61% ($486M) organic chemicals share of the market. The 1981 US market analysis, Figure 1, follows a similar chemicals make up, citing amines (e.g. fatty acid amines, cyclic amines, quaternaries, diethylamines and imidazolines) used predominantly as corrosion inhibitors in oil and gas production/ transmission, refineries and boiler water return lines. In the case of cooling water treatment 95%

US CORROSION INHIBITOR MARKET

1981		1986
$573M	5.5% per annum -------------> **TOTAL**	$750M

END USE MARKET

<u>Process side</u>
e.g. oil production
refining
pipelines

51.9%	7% per annum ----------->	55.2%
50.4%	oil production ------------->	53.4%
26%	refining ------->	24%

<u>Water side</u>
e.g. cooling water operations
boiler water operations

48.1%	3.8% per annum ------------->	44.8%
37.8%	chemical/petrochemical --------------------->	38.6%
25.1%	refinery ------->	23.6%

Figure 1

of corrosion inhibitors used were based on chromates, zinc salts, phosphates, molybdates and triazoles. Phosphonates and some polymers are identified then as new corrosion inhibitor bases; phosphonates feature in the 1986 survey.

The EEC market analysis, Figure 2, would suggest that inorganic based corrosion inhibitors here take the major percentage share reflecting the dominance of water treatment. However, an increasing use of organic corrosion inhibitors is predicted for treatment of cooling waters due to a decline in the use of heavy metals (e.g. Cr, Zn). An increased use of neutralising amines for treatment of boiler water return lines and a decrease in use of oxygen scavengers to treat boiler waters is also seen as significant in shifting the balance towards greater use of organic chemicals. An interesting aspect of the figures is the size of the anti-freeze market, falling not far behind that for oil production and refining. Corrosion inhibitors typically used with anti-freeze formulations include sodium benzoate, borax, benzotriazole and tolytriazole. However, with the automotive industry's move to multi-metal engines inhibitor formulations are becoming more complex.

3 DEFINITION OF A CORROSION INHIBITOR

At the simplest level a corrosion inhibitor is a substance which when added in small concentration to an environment effectively reduces the corrosion rate of a metal exposed to that environment. A more scientifically exacting definition is really not possible in the general sense where mechanistic and/or chemical considerations serve to sub-divide corrosion inhibitors into specific types or classes.

In the majority of cases inhibition is achieved through interaction or reaction between the corrosion inhibitor and the metal surface, resulting in the formation of an inhibitive surface film; this may occur directly at the metal/environment interface or via precipitation. In other cases the chemistry of the environment may be modified to render it less corrosive, e.g. by chemically scavenging dissolved oxygen; chemically neutralising dissolved acidic gases; adjusting pH to promote stable (inherent) passivation.

EEC CORROSION INHIBITOR MARKET

1980		1990
%122.3M (81 000 tons)	4% per annum ------------>	$180M (117 000 tons)
	TOTAL	
63 0000 tons (78%)	3.4% per annum --------------> Water Treatment	88 000 tons (75%)
3 000 tons (3.7%)	11% - 7% per annum -----------------> Oil Well & Refinery	7 000 tons (6.0%)
4 600 tons (5.7%)	3 - 3.5% per annum -----------------> Antifreeze	6 100 tons (5.2%)
	LARGEST OUTLETS	
$33.7M	------------> West Germany	$48.3M
$24.5M	--------------> United Kingdom	$40.5M
$23.9M	-----> France	$34M

Figure 2

TABLE 1

TYPE	SPECIES	FURTHER CLASSIFICATION
	Chromate	Oxidising Passivator
Anodic[1]	Nitrite	Oxidising Passivator
	Nitrite	Oxidising Passivator
	Phosphate	Non-Oxidising Passivator
Anodic[1]	Molybdate	Non-Oxidising Passivator
	Tungstate	Non-Oxidising Passivator
Cathodic	Arsenates	Cathodic Poison
Cathodic[2]	Carbonates	Precipitates (Ca & Mg)
Cathodic[2]	Zn Sulphate	Precipitate as $Zn(OH)_2$
Precipitation	Phosphates	Anodic/Cathodic Effects
Precipitation	Silicates	Anodic/Cathodic Effects
O_2 Scavenger	Sulphite	Cathodic in effect
O_2 Scavenger	Hydrazine	Cathodic in effect
Volatile/	Cyclohexylamines	Passivating (NO_2^- salt)
Vapour Phase	Morpholine	Neutralising
	Amines	Organic/Adsorption
	Amines	Organic/Adsorption
Film-Forming[3]	Imidazolines	Organic/Adsorption
	Quaternaries	Organic/Adsorption
	Acetylenic Alcohols	Organic/Adsorption

[1] Sometimes referred to as 'dangerous' inhibitors because of a need to be present above critical concentrations if the promotion of pitting corrosion is to be avoided, i.e. to ensure all the active anodic sites/areas are inhibited.

[2] Sometimes referred to as 'safe' inhibitors because they do not directly interfere with the anodic reaction but reduce the net current flowing in the overally corrosion reaction as a consequence of inhibiting the cathodic reaction.

[3] May well exhibit anodic, cathodic or mixed behaviour (i.e. a combination of both anodic and cathodic ettects).

Inhibited paint films may seem a special case, but really the aforementioned considerations still apply in general, but now being affected by how readily inhibitor is released (i.e. inhibitor solubility) into the corrosive environment as it moves through the paint film towards the metal substrate. Recent developments using inhibitor-loaded ion-exchange pigments[3] would appear to introduce an added consideration however. Here the inhibitor is released off the supporting pigment by an ion-exchange mechanism at a rate dependent on the corrosive properties of its surroundings; the implication then being that inhibitor is released in a more controlled manner in response to the corrosivity of the invading environment.

4 CLASSIFICATION OF CORROSION INHIBITORS

Classification of corrosion inhibitors is somewhat a subjective exercise being very much dependent on the method employed. For example, one could choose to classify in terms of mechanism, environment to which they are added, metal to be protected or inhibitor chemistry, of which none are mutually exclusive. Some of the more commonly encountered descriptives are anodic, cathodic, passivating, oxidising, film-forming, organic, vapour phase, volatile, and 'safe' or 'dangerous' inhibitors. Examples of the aforementioned are given in Table 1.

Electrochemistry. Aqueous systems are by far the most common corrosive environments to which corrosion inhibitors are applied. It is then particularly informative to examine how a corrosion inhibitor affects the electrochemical reactions which combine to describe the corrosion process to be inhibited.

Mixed potential theory of an electrochemical corrosion process expressed graphically in the form of an Evans type polarisation diagram is an extremely useful way to picture how an inhibitor works. Figure 3 shows a simple corrosion reaction described in this manner defining the steady state free corrosion potential and corrosion current (rate). These values result from the mutually polarising (coupled) partial (or half) reactions, anodic and cathodic, which combine to constitute the overall corrosion reaction. Figure 4 shows 6 basic changes that can result from the situation described in Figure 3 as a result of adding a corrosion inhibitor[4]. It can be seen that a corrosion inhibitor can either affect (or interfere

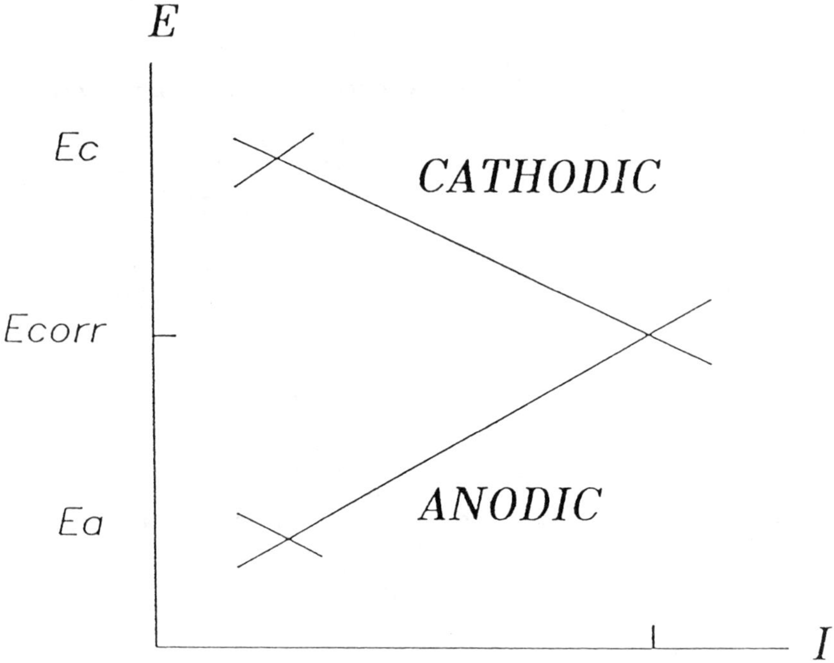

ANODIC $M \longrightarrow M^{n+} + ne^-$

CATHODIC <u>Acid</u>

$2H^+ + 2e^- \longrightarrow H_2$

$O_2 + 4H^+ + 4e^- \longrightarrow 2H_2O$

<u>Neutral</u> <u>to</u> <u>Alkaline</u>

$2H_2O + 2e^- \longrightarrow 2OH^- + H_2$

$O_2 + 2H_2O + 4e^- \longrightarrow 4OH^-$

Figure 3. Evans type polarisation diagram describing a simple corrosion process.

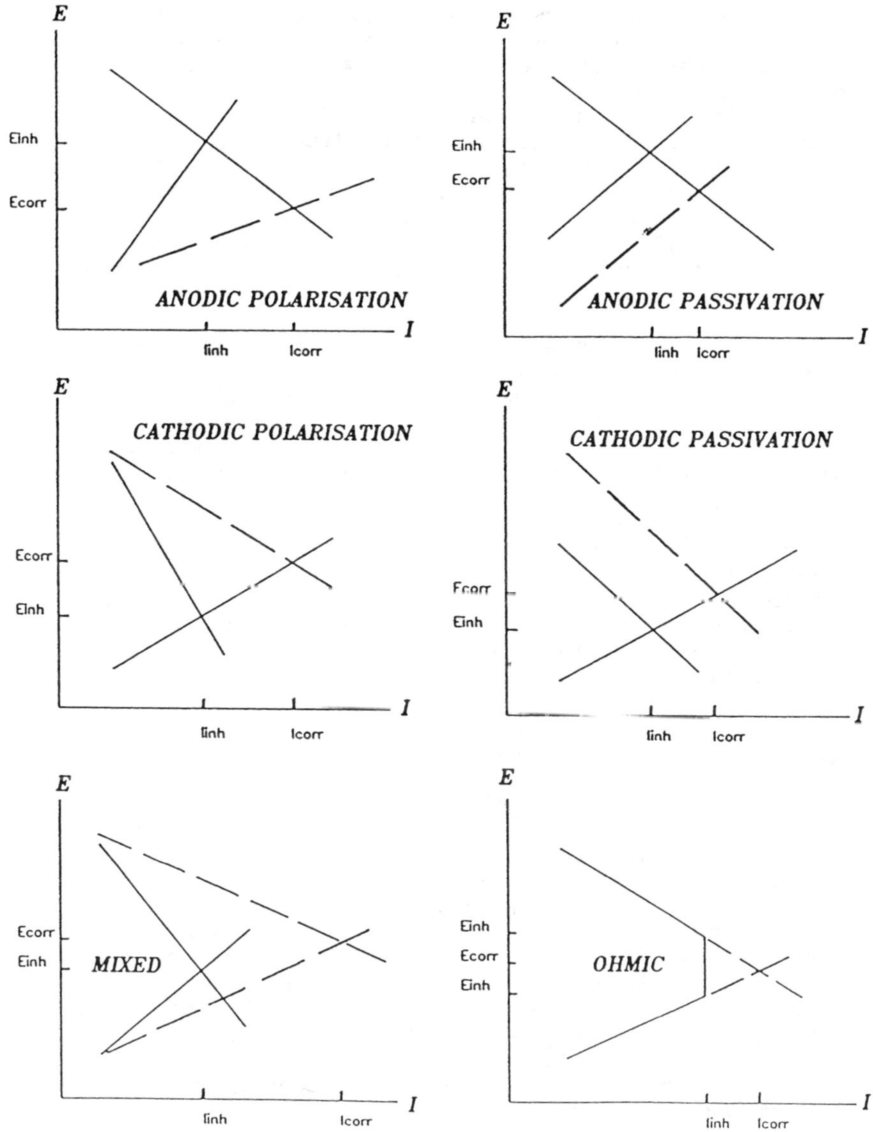

Figure 4. Evans type polarisation diagram describing how a chemical inhibitor can effect corrosion inhibition.

with) the anodic partial reaction, cathodic partial reaction or both. Immediately one can see where the classifications of anodic and cathodic inhibitors derive: in reducing the corrosion current the corrosion potential is shifted anodically (positively) for an anodic inhibitor and cathodically (negatively) for a cathodic inhibitor. In the case where the inhibitor affects both partial reactions, often a common feature in practice, the shift in corrosion potential will depend on which is the more dominant effect and little or no potential shift may in fact occur.

The remaining situation pictured in Figure 4 is the presence of an ohmic resistance. This may result naturally from the presence of a potential drop in solution or in the presence of a poorly conducting film, a feature associated with certain types of inhibitor films (e.g. inhibition of copper and copper alloys by benzotriazole) and paint films.

Passivation From the foregoing it is then possible to describe more specific inhibitor types. Passivating inhibitors are a specific type of anodic inhibitor. They cause a large potential shift in corrosion potential to render a metal passive, and if present in sufficient concentration are generally the most effective of all inhibitors because they can stifle corrosion almost completely. However, the emphasis lies on their presence at a sufficient (critical) concentration, otherwise they can promote localised corrosion through being unable to inhibit all the active anodic sites (areas); and are therefore often classed as dangerous inhibitors. Conversely, cathodic inhibitors are often classed as safe inhibitors as they do not interact directly with the anodic reaction. Nevertheless there are notable exceptions in the form of cathodic poisons, which interfere with the cathodic hydrogen recombination reaction so reducing general corrosion, but can promote hydrogen blistering and embrittlement due to absorption of atomic hydrogen by the metal.

The picture for a passivating inhibitor may be generally described as in Figure 5, where if the inhibitor is present below a critical concentration the corrosion potential can sit either in the active or passive region, an unstable situation generally resulting in pitting. However, in drawing parallels

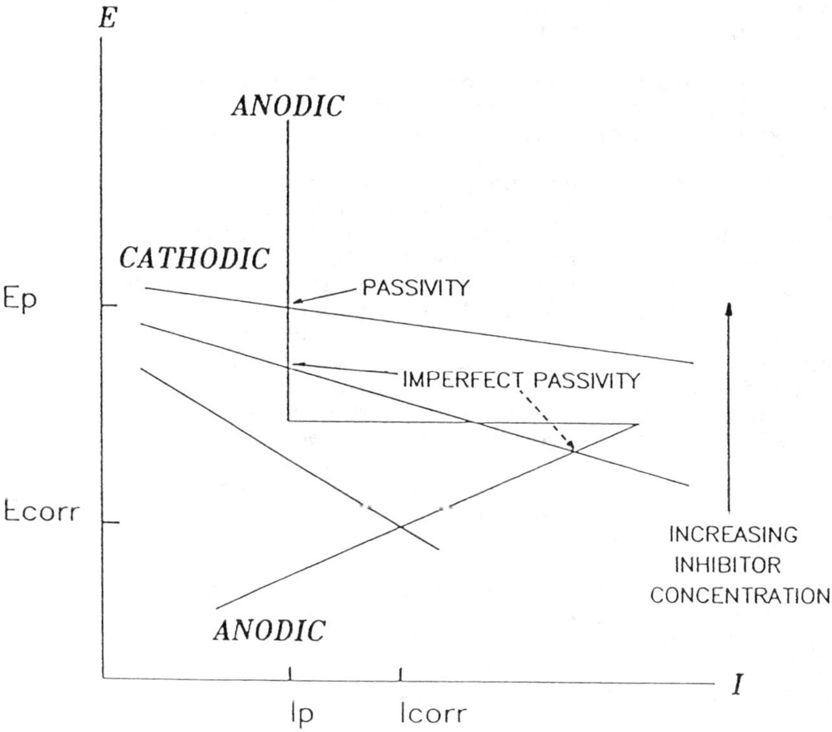

Figure 5. Evans type polarisation describing the action of a passivating anodic corrosion inhibitor.

with the simple picture of an anodic inhibitor in Figure 4, there are some important differences. The basic criterion of inhibition resulting from an anodic shift in corrosion potential certainly applies, but just how is this achieved?

The polarisation diagram of Figure 5 shows no basic change in the shape/position of the overall anodic polarisation curve, whereas that for the cathodic reaction changes significantly! The inhibitor is causing the potential of the corroding metal to be shifted sufficiently anodic for it to sit within the thermodynamically defined passive region inherent to that metal. The inhibitor is thus oxidising in its action on the metal and as such the degree of this effect will be dependent on inhibitor concentration. Typical oxidising inhibitors are chromate and nitrite which are themselves easily reduced and so have the effect of depolarising the cathodic reaction, as is shown in Figure 5. However, certainly in the case of chromate, adsorption of chromate at anodic areas also appears to play a part in the process, polarising the anodic reaction and so decreasing the anodic current required to move the potential into the passive region.

Certain passivating inhibitors, such as phosphate, silicate and molybdate, are classed as non-oxidising as they require the presence of oxygen to induce passivity; however, they still can be dangerous if present in insufficient concentration.

Thus in reviewing this example, inhibitor classification moved through passivating to oxidising, dangerous and non-oxidising with some help from cathodic depolarisation and adsorption whilst still fulfilling the basic classification of anodic. Additionally the inhibitor anion usually gets incorporated into the passive oxide film formed further helping to stabilise it. Such is the difficulty of unambiguous classification, although the underlying principles can be defined.

<u>Adsorption</u> Adsorption is an important step in inhibition by organic-based film-forming corrosion inhibitors. Adsorption results from the polar or charged nature of the organic molecule/ionic species first establishing a physisorbed surface film (through van der Waals forces) which may further stabilise through chemisorption to form a donor type bond

(Figure 6). The latter has been considered in terms of Lewis acid and base theory applied to the stability of the donor (adsorption) bond; the Hard and Soft Acids and Bases (HASB) principle[5] where a hard acid will bond with a hard base and vice versa. The principle is exemplified in Figure 7.

Adsorption is the primary step in achieving inhibition in acid solutions. This is a consequence of the fact that the corroding metal surface to be inhibited is usually oxide-free allowing the inhibitor ready access to retard the cathodic and/or the anodic electrochemical processes of corrosion. Once the inhibitor has adsorbed on the metal surface it can then affect the corrosion reactions in a number of ways[6]: by offering a physical barrier to the diffusion of ions or molecules to or from the metal surface; direct blocking of anodic and/or cathodic reaction sites; interaction with adsorbed corrosion reaction intermediates; change the make-up of the electrical double layer which develops at the metal/solution interface and so affect the rate of electrochemical reactions taking place there.

Precipitation In near-neutral solutions metal surfaces are usually covered with oxides, hydroxides or salts present due to their reduced solubility in the region of the corroding metal/solution interface. Here an inhibitor serves to stabilise the situation through forming a protective surface film; a passivating action. The act of precipitation alone can serve to inhibit corrosion, the precipitated film presenting a barrier to diffusion of oxygen to the metal surface. The implication here is that performance is associated with the involvement of oxygen in the overall corrosion process and as such inhibitors are often classed as cathodic inhibitors. The fact that the pH local to the cathodic oxygen reduction reaction tends to favour precipitation preferentially, at cathodic areas, of many salts effecting such inhibition also leads to the cathodic inhibitor classification. The most common salts in this class are insoluble hydroxides of zinc and magnesium and carbonates of calcium and magnesium, where the inhibition can be directly initiated by adjusting the bulk pH.

In practice such precipitated films will tend to generally cover a metal surface so that they may also exhibit some anodic character. In the case of phosphates and silicates earlier classed as

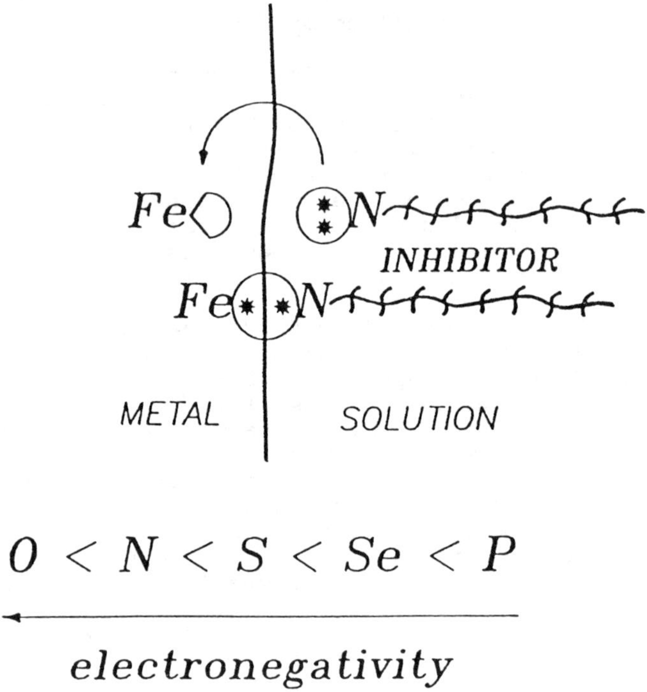

Figure 6. Schematic representation of the chemisorption process for an organic film-forming corrosion inhibitor.

HARD	**low polarisability** **high electronegativity**
SOFT	**high polarisability** **low electronegativity**

HARD ACID	**HARD BASE**
iron oxide	**O, N**
SOFT ACID	**SOFT BASE**
active iron	**S, P, Se, As**

Figure 7. The Hard and soft Acid and Base (HSAB) principle.

non-oxidising passivators, precipitation may well play a significant role with inhibitor action appearing to be mixed.

5 CORROSION INHIBITORS IN PRACTICE

The papers that follow look in more detail at specific applications of corrosion inhibitors and inhibitor testing/selection, so comment here is restricted to some general observations on practice.

The range of corrosion inhibitors available is large which in itself can make selection often appear a daunting task. This is particularly true with regard to the ever growing range of organic based corrosion inhibitors, where small changes in inhibitor chemistry can have significant effects on performance. Inhibitor performance, and thus selection, will be affected by the nature of the metal and that of its surface, nature and composition of the environment, temperature, fluid dynamics and inhibitor concentration/treatment. Put on top of this the presence of dissimilar metals in the same system, presence of crevices, other surface deposits (scale), presence of other treating chemicals (e.g. scale inhibitor) and toxicity regulations, and inhibitor selection becomes far from being a straightforward choice. All this serves to highlight the need to understand, and ideally accurately describe, the system to be inhibited, and the need to conduct laboratory studies and testing if the right inhibitor is to be finally selected and used in a cost effective manner.

Electrochemical test methods are now very much to the fore in the study of corrosion inhibitors. The knowledge and understanding of corrosion inhibitors has benefited enormously from electrochemical studies, but the simple weight loss/exposure type test still has a reassuring feel to it for many; even if used only as a final confirmation. Nevertheless, organic coatings, where the norm is still exposure testing, are now benefiting from electrochemical measurements, notably AC impedance[7]. However, in looking at test methods no one method will necessarily provide all the answers from which final inhibitor selection can be safely made[8]. It is important to understand just what a particular test method is actually measuring, and its limitations, and the implications of this in response to possible mechanisms of inhibition. For example, work by Hausler[9] on high temperature CO_2 corrosion of carbon steel points to inhibition being achieved by retardation of the dissolution of an iron carbonate scale layer: such an effect cannot be observed by electrochemical measurement in constant inventory autoclave tests.

Although the application of corrosion inhibitors is broad based, their principle use lies in the inhibition of general corrosion. Localised corrosion, in its many forms, presents added problems to the successful application of a corrosion inhibitor. This may in part be associated with the occluded corrosion cell that often results from restricting inhibitor access, of particular importance to an anodic inhibitor. However, as highlighted earlier, cathodic inhibitors can have their problems (cf cathodic poisons) which may promote cracking due to hydrogen damage. Such situations do require even more careful consideration but nevertheless examples of inhibition of crevice corrosion and pitting[10], and stress corrosion[11] have been reported. Professor Parkins' concluding remarks on inhibition of stress corrosion cracking (Cavallaro Medal Lecture - 6th European Symposium on Corrosion Inhibitors[12]) are perhaps salutary to the inhibition of localised corrosion per se viz. '..... if attempts are made to apply experiences in relation to general corrosion to the solution of environment related cracking ..... such an approach will be frequently ineffective, which is perhaps not surprising since many of the environments that promote environment sensitive fracture are often those environments formulated to prevent general corrosion.'

As a final comment here, dependent on the application, a commercial corrosion inhibitor product may in fact be a balanced blend of actives or an active(s) formulated with a specific surfactant/ solvent package. This brings into play the word synergism. The inhibition of mild steel by zinc-molybdate mixtures is an example of synergism, where the corrosion rate is lower in the presence of both than in solutions of the individual components[13]. Inhibition is achieved as a result of the two ions combining to retard the cathodic reduction of oxygen. It is possible to produce somewhat similar effects on corrosion rate with organic film-forming corrosion inhibitors of the type common to oilfield applications (e.g. imidazolines): the effects may be achieved by changing/removing the surfactant or changing the solvent package. Here it is questionable whether this can be classed as synergism but is more to do with changing/modifying the physicochemistry, e.g. the surfactant makes the metal surface more receptive to adsorption by the corrosion inhibitor species whereas the solvent package affects partitioning and/or dispersibility of the inhibitor (i.e. affects available/active inhibitor concentration in solution). Clearly blending and formulating are added complexities, but equally may be seen as another level of sophistication being brought to bear on the development of commercial corrosion inhibitor products; a natural step forward as the mechanisms by which corrosion inhibitors work become better understood.

6 SUMMARY

Corrosion inhibitors have a continuing and major role to play in the cost effective control of corrosion. Their application is broad based and extensive representing a mature technology in many areas.

Basic understanding of inhibition has benefited significantly from the application of modern electrochemical techniques and theory to better describe inhibitor performance. This has helped in promoting an increasing level of sophistication in the development of commercial corrosion inhibitor products and practice.

The inhibitor market appears to be following predicted growth figures (4 - 5% over the period 1981 - 90) with the current total market value probably sitting at about $1B. Organic-based chemical inhibitors appear to have the majority share of the market, and it will be interesting to see how strongly this penetration continues. This may be assisted by increasingly tighter environmental regulations which are likely to further question the continuing use and acceptability of the more toxic inhibitors such as zinc chromate.

BIBLIOGRAPHY

'Corrosion Inhibitors', J.I. Bregman, The Macmillan Co., New York (1963)

'Mechanism and Phenomenology of Organic Inhibitors',

G. Trabanelli & V. Carassiti, 'Advances in Corrosion Science and Technology' (Volume 1), Ed. M.G. Fontanta & R.W. Staechle, Plenum Press, New York (1970)

'Corrosion Inhibition: Principles and Practice', A.D. Mercer 'Corrosion' (Volume 2), Ed. L.L. Shreir, Newnes-Butterworths, London (1976)

'The Mechanism of Corrosion Prevention by Inhibitors', J.G.N. Thomas, 'Corrosion' (Volume 2), Ed. L.L. Shreir, Newnes-Butterworths, London (1976)

'Corrosion Inhibitors', Ed. C.C. Nathan, NACE, Houston, Texas (1974)

ACKNOWLEDGEMENTS

The author would like to thank the British Petroleum Company plc, for permission to present this paper.

REFERENCES

1. 'US Corrosion Inhibitor Market Study' Hochberg & Co, Chester, New Jersey (report in 'Chemical Market Reporter', Schnell Publishing Corporation Inc., New York, July 1987)

2. 'Corrosion Inhibitors Market in Europe' and 'Corrosion Inhibitors Market in the US', Frost & Sullivan[2] Inc., New York (reviewed in *Anticorrosion,* May and August 1982)

3. B.P.F. Goldie *Polymers Paint Colour J,* 1985, 5,337

4. G. Bailey & M.E.D. Turner, *Industrial Corrosion,* 4(4), 9, 1986

5. R.G. Pearson, *J. Chemical Ed.,* 45(9), 581, 1968

6. T.G.N. Thomas, 'Corrosion' (Volume 2), Ed. L.L. Shreir, Newnes-Butterworths, London 1976 18:38

7. G.W. Walter, *Corrosion Sci.*, 26(9), 681, 1986

8. D. Harrop, 'Proceedings of UK Corrosion 87', Brighton, England, 1987

9. R.H. Hausler, 'Proceedings of 6th European Symposium on Corrosion Inhibitors', Ferrara, Italy, 1, 41, 1985

10. E. McCafferty 'Proceedings of 6th European Symposium on Corrosion Inhibitors', Ferrara, Italy, 1, 533, 1985

11. C.M. Rangel & R.N. Parkins 'Proceedings of 6th European Symposium on Corrosion Inhibitors', Ferrara, Italy, 1, 655, 1985

12. R.N. Parkins, 'Proceedings of 6th European Symposium on Corrosion Inhibitors', Ferrara, Italy, 1, 1, 1985

13. Y.J. Qian & S. Turgoose, *Br. Corrosion J.*, 22(4) 268, 1987

The Design of Corrosion Inhibitors to Meet Specific Requirements for Oilfield Applications

By J. E. Donham

DONHAM INTERNATIONAL SERVICE COMPANY INCORPORATED

1. INTRODUCTION

One major method of corrosion control is the use of inhibitors in drilling and stimulation fluids, producing wells, water injection systems, transportation, and refining systems. Inhibitors are also applied externally to prevent rusting of stored tubular goods.

Inhibitors are synthesised from a relatively few basic materials and depend upon formation of a chemically adsorbed film, passivation of a metal surface by altering the corrosion byproduct formed, or forming a complex with the metal surface.

The end use requirements must be the guide for inhibitor development. These requirements depend upon the corrosive environment and corrodents, the metal to be protected, and equipment design parameters.

The ability of the inhibitor to form an impervious, long-lasting film depends upon its chemical structure and interaction with the corrosive environment. This paper addresses the synthesis, formulation, testing, and proper application of inhibitors based upon the requirements of particular systems in oil production and transportation.

Monitoring to ensure that the inhibitor is meeting the requirements, downstream effects of the inhibitor on oil-water separation, compatibility with other treating chemicals and processes, environmental considerations, and effects of microbial activity will also be considered.

2. DEFINING THE REQUIREMENTS

Before a corrosion prevention procedure using inhibitors can be implemented, the system requirements must be clearly understood. The following problem areas and parameters will dictate the requirements and performance specifications of a particular inhibitor.

PIPELINES AND FLOWLINES

Top of Pipe. The 12 o'clock position in a line is the most difficult to inhibit. Where the flow velocity is less than required for turbulent flow, liquids will not contact this area except in areas of slug or partial slug flow. Addition of a volatile component to the inhibitor may be required.

Water Wet Area-Bottom of Pipe. In most cases free liquid moves along the bottom of the pipe. Depending upon velocity, the layer may be both condensate and water, or as discrete layers of oil and water.

At low velocities solids dropout can cause concentration cells and pits underneath the deposits.

Turbulence Prone Areas. Areas downstream of welds, minor buckling of the line, low spots and solids deposits can increase shear stress and turbulence which may aggravate corrosion. Low spots cause slugs of liquid at intervals.

Turbulence removes protective scale, aggravates abrasion and erosion if solids are present, and may affect inhibitor performance by removing the film. Inhibitors must be able to withstand the shear stresses.

PRODUCING SYSTEMS

Tubing. Protection of production tubing requires that the inhibitor be squeezed, added continuously or have film persistency so that batch treatment is feasible. Shear stresses impose the same requirements on inhibitors to withstand velocity effects. The tubing will be wetted more completely on the low side in deviated (non-vertical) wells, and in low-volume producers a separate layer of water on the low side may be the only corrosive area.

Downhole Pumps. In rod pumped wells abrasion of rods on tubing adds the requirement that the inhibitor film have some lubricating properties. In many wells entry of air through the annulus requires that the inhibitor function in the presence of oxygen.

Downhole centrifugal pumps are subsceptible to failure due to scaling, and to velocity effects.

Surface Equipment. Problem areas include wellheads, chokes, and vessels where velocity effects are substantial, and separators and other vessels where stagnant areas, scale and solids deposits can create concentration cells. Growth of bacteria may occur in these areas.

Injection Systems. Secondary recovery (waterflooding) and tertiary recovery using CO_2, micellar fluids, and polymers introduce some specific requirements. Inhibitors must have proper solubility and wetting characteristics and be able to perform in the presence of the surfactants and polymers added to the flood. A surfactant component may be required in the inhibitor to maintain injection rates in produced water injection systems.

GAS/LIQUID COMPOSITION AND OPERATING CONDITIONS

The major controlling factors for corrosion rates are the composition of the gases and liquids produced or transported, and conditions of flow, temperature, and pressure.

H_2S-CO_2 Content. The acid gas content determines the type of corrosion and greatly influences the corrosion rate. Corrosion rates are directly related to the amount of CO_2 dissolved in the water which determines the amount of carbonic acid and subsequent metal dissolution. If organic acids are present corrosion rate is increased by removal of bicarbonate ion, and by dissolution of protective iron carbonate:

$$CH_3COOH + HCO_3^- \longrightarrow CH_3COO^- + CO_2 + H_2O$$

$$2CH_3COOH + FeCO_3 \longrightarrow Fe(CH_3COO)_2 + 2H^+$$

High velocity causes turbulence and increases corrosion rates. The temperature of the system, increased salinity and bicarbonate content also affect the corrosion rate and the inhibitor requirements.

Where H_2S is present line failure due to penetration underneath pits can occur in a short time. The sulphide film formed may be anodic to the metal surface, and afford some degree of corrosion protection. In many cases, however, the layer of FeS is not continuous and if so, may be porous. The net result is pit formation and growth.

Sulphide stress cracking and hydrogen embrittlement are also factors to consider in inhibitors for H_2S.

In systems where H_2S and CO_2 are both present, the ratio of CO_2 to H_2S determines whether CO_2 or H_2S corrosion mechanisms will dominate.

Inhibitors must be effective against H_2S and CO_2.

Liquid Composition. Water composition may range from water of condensation to high salinity formation water. Inhibitor solubility and dispersibility requirements will be affected.

Water floods range from sea water to mixtures of re-injected produced water with surface and well water. In some cases fresh produced water is discharged into streams so environmental requirements are of concern.

Hydrocarbons may range from low molecular weight aliphatics to high molecular weight asphaltenes. Condensate usually does not enhance inhibitor filming, while higher molecular weight hydrocarbons and organic compounds may help the inhibitor. Paraffins and waxes form deposits and require treatment.

Temperature. Carbon steel corrosion by CO_2 is directly dependent upon temperature. At temperatures below 60°C scale provides little corrosion protection. In the temperature range of 60 to 100°C iron carbonate will form both remote and on the surface. The scale may result in lower than predicted corrosion rates, but severe pitting can (and usually does) occur.

At temperatures above 100°C scale is formed on the surface as a thin, dense layer of iron carbonate/

magnetite, and affords good corrosion protection under most conditions.

Temperature affects the inhibitor requirements as far as stability and filming ability is concerned. No polymerization of the inhibitor can be tolerated, since plugging can take place.

Pressure. Pressure has a direct effect since corrosion rates are proportional to acid gas partial pressures. It will affect inhibitor solubility since at extremely high pressures methane will act like a liquid and may remove the solvent from the inhibitor, leaving a thick residue.

Flow Parameters. The flow parameters to consider are velocity, type of flow, and gas to liquids ratio.

Velocity has definite effects on the ability of inhibitors to control corrosion. The type of flow is determined by velocity, is characterised as annular, stratified, or slug flow. Flow regimes vary in different sections of lines and tubing, because of restrictions, and low places. Slug flow in producing wells increases turbulence.

Distribution of the inhibitor in all areas of a pipe line is related to flow velocities and the composition of the gas and liquids in the line. At annular flow velocities, the stream is homogeneous, and inhibitor added continuously will contact all portions of the line equally.

When flow velocities are lower, the flow is partly annular, with a higher concentration of liquids in the bottom half of the pipe- i.e. the film of liquids is thicker. Lower flow velocities allow some free liquid to collect in the bottom of the line, and slug flow predominates. As the flow rate declines stratified flow predominates and the stream is a gas phase and a liquid phase. The top portion may not be regularly contacted at all with liquid except that condensed from the gas. A vapor phase inhibitor or some means of a periodic batch that contacts the top of the line must be considered.

In stratified flow, partitioning of the inhibitor between oil and water layers is important. Slug flow increases requirements for ability to withstand shear stress.

At velocities of less than 10 m/s it is reported that very little effect on CO_2 corrosion rates will take place. At velocities of 10 to 20 m/s, turbulence can cause local areas of higher attack, and at velocities above 20 m/s corrosion byproducts will be removed. This increases corrosion rates and could affect inhibitor filming ability.

OTHER FACTORS AFFECTING CORROSION INHIBITOR REQUIREMENTS

Other factors affecting corrosion include bacteria, scaling, mechanical or chemical treatment of lines prior to commissioning such as treatment of the pipe during storage, and completion methods and treating procedures. Economics of treating is important. Compatibility of the inhibitor with scale inhibitors, oxygen scavengers, and biocides imposes special requirements.

Bacteria. Bacteria, particularly sulphate reducers, can increase corrosion rates and may need to be controlled with organic biocides.

Organisms may be introduced during hydrotesting of lines, and by contamination of producing systems from sumps and sea water flushing of vessels.

Scale. Scaling may need to be controlled to prevent pitting corrosion. The inhibitor may be a combination material or scale inhibitor may be used separately.

Mechanical and Chemical Pretreatment. Pipelines may require treatment to remove mill scale and deposits prior to a successful inhibitor treatment regime, and will be hydrotested. Prevention of corrosion and bacterial growth during testing is required.

Completion Methods and Treating Procedures. Completion methods dictate the treating procedures. If the well is flowing with a packed-off annulus it may be necessary to install a chemical string for continuous treatment. Side pocket mandrels with chemical injection valves or capillary strings require that inhibitors be stable in the annulus or the string for extended periods.

If a chemical string is not feasible, batch treatments using film persistent inhibitors may be used. The inhibitor is designed to form a tough film that is not too soluble in the production stream so it will last for a sufficient time between treatment. The batch may be displaced with liquids, gas, or nitrogen.

Squeeze inhibitors must be designed to be stable in the formation, and not cause severe emulsion problems in the formation. The adsorption characteristics must be controlled for proper feed back of inhibitor.

Pumped wells can be treated by continuous addition or batching down the annulus.

Economics. Economics of inhibitor use is an important requirement and includes inhibitor costs, treating level any non-routine treating procedures, monitoring, volumes of water to be treated, and development costs for these factors.

Compatibility with other Chemicals. The inhibitor must be able to treat in the presence of other materials such as phosphonates, polymers, bisulphites, and surface active agents, and must not interfere with their functions.

3. DESISIGNING THE INHIBITORS.

TEST METHODS AND MONITORING.

Proper inhibitor evaluation and development depends upon selecting a test method that will examine particular requirements, such as film strength, or is within parameters that stimulate actual field conditions. An example of the first is the bubble test, and an example of the second is a flowing loop test.

Corrosion measurements are made by weight loss, linear polarization, potentiodynamic measurement, alternating current impedance, and electrical noise.

Some of the test methods used for inhibitor selection are listed below.

Static Tests. This involves placing a coupon in a corrosive environment and leaving it quiescent for a period of time. The coupon is then inspected, pitting noted, and cleaned and weighed to determine weight loss and per cent protection. The test can be run at different temperatures and pressures, and continuous and filmed inhibition can be tested.

Stirred Flask (Bubble) Test. This involves bubbling CO_2 (or mixed gases or H_2S) through a stirred flask maintained at a specified temperature. The corrosion rate is read with electrodes immersed in the flask. Oil-water mixtures can be varied to test the partitioning of the inhibitor in the oil and water. Filming speed and the level of inhibition can be determined by LPR, AC impedance, electrochemical noise measurements and weight loss coupons.

Autoclave Test. Autoclave tests may be stirred, agitated, or static. High pressures and temperatures may be obtained. These tests impose more severe conditions on the inhibitors under test, and their stability under extreme conditions can be evaluated.

Rotating Cylinder or Electrode Tests. A flask containing a rotating electrode assembly can test the effect of shear stress on corrosion rates and inhibitor film strength. The shear stress can be calculated and turbulent flow in a pipeline at known hydrodynamic conditions simulated.

A cylindrical test specimen is attached to the bottom of a shaft in a standard rotating disk apparatus with a variable speed motor. A slip ring assembly is used for the electrical contact to the rotating electrode.

The effect of different shear stresses and inhibitor addition is observed and recorded. LPR, electrical noise, and AC impedance measurements may be used to measure corrosion rates.

One variation of this test uses a stationary electrode and a rotating cylinder.

Flowing Tests. This test simulates flow conditions found in a pipeline. The tests use weight loss coupons, electrochemical measurements, and analysis of the circulating media. The fluids, acid gases, temperature, and velocities can be altered as required.

The test can be modified for jet impingement studies. Jet impingement on a weight loss coupon or microelectrode assembly simulates high turbulence and mixed flow conditions. There will be a central laminar flow region, a transition zone with maximum shear stress, and an outer zone where the turbulence decays proportionally to the distance from the central region.

Wheel Tests. Coupons are placed in bottles or flasks containing water or water and oil. The bottles are purged free of oxygen, saturated with the acid gas to be tested, sealed and installed on a rotating wheel or cylinder in a heated chamber for a specified period of time. The coupons are removed, inspected, and weighed. Pitting and blistering is noted and the tendency of the inhibitor to form a stable emulsion with the oil in the test is also noted and recorded.

Inhibitor can be added to the bottles, or the coupons can be filmed before being put in the bottles.

TYPES OF CHEMICALS AND INHIBITOR MECHANISMS

The types of chemicals used in inhibitors are predominantly fatty nitrogen compounds or organic amines. They will be materials that either have quick filming and high resistance to wash-off, or have extremely long film persistency. They should be selected to be effective against H_2S, CO_2 and organic acids in particular. In some cases resistance to oxygen contamination is important.

They should exhibit at least 90 per cent protection on the tests selected for final evaluation.

Organic amine inhibitors function by forming a chemisorption bond with a metal surface. The nitrogens in the amine are cationic and bond with anionic sites on the metal surface. There is evidence to suggest that the amine forms a complex with oxides and sulphides on the metal, and perhaps with the metal itself. The mechanism is quite complicated, and there is some evidence that the molecule actually re-arranges and forms a polymeric material on the surface. Some work indicates that the nitrogen acts as the "leader" to the surface, and the polymeric complex forms a strong bond to the metal. The bond is particularly strong to a sulphide layer but amines do not bond strongly to an oxide layer. The presence of oxygen interferes with their performance unless additives are formulated in the inhibitor blend to counteract this.

The filming effect is enhanced by the presence of materials such as fatty acids and naturally occurring high molecular weight organic acids in petroleum. Addition of polybasic acids to the inhibitor formulation can cause cross-bonding of the amine layer, and increase film life or persistency. Certain types of organic acids and organic compounds containing phosphorus, sulphur, and other elements increase the inhibition in the presence of oxygen.

The layer of inhibitor isolates the metal from the electrolyte and stops the corrosion reaction. In continuous treatment, the film is replenished by inhibitor in solution. For batch treating, an amine or amine blend is selected that forms a long-lasting film that is replenished at intervals by another batch of inhibitor.

INHIBITOR RAW MATERIALS.

The basic raw materials for preparing inhibitors are fatty acids(or synthetic long chain aliphatic acids), naphthenic acids, oxidised petroleum fractions, and ammonia or organic amines. Acrylonitrile and alpha olefins are also used for synthesising amines.
Attachment 1 shows some of these chemicals and their reactions.

Fatty Acids. These range from low molecular weight organic acids such as acetic acid up to C_{18} and higher. The low molecular weight acids are normally used only to salt amines to achieve water solubility or dispersibility, while the higher molecular weight acids are used to form fatty amines. Fatty acids based on vegetable and animal fats such as tallow, coconut oil, and tall oil are the most commonly used and include lauric, stearic, oleic, linoleic, and linolenic acids. Unsaturated fatty acids may be polymerised to form polycarboxylic acids.

Amines. These range from ammonia (NH_3) to high molecular weight ethyleneamines. One manufacturing process for making ethylene amines is the reaction between ethylene dichloride and ammonia.

The reaction product, ethylenediamine, can be used to make imidazolines and amido-amines. It may be further reacted with ethylene dichloride to form diethylenetriamine and higher molecular weight ethyleneamines.

Fatty Amines. Primary fatty amines are made by reacting fatty acids with ammonia to form amides, then hydrogenating to the amine. Primary amines can also be made from alpha olefins.

Secondary and tertiary amines are also synthesised from alpha olefins and may be further reacted to make amides, quaternaries, and the like.

Fatty Diamines. These are synthesised by reacting a primary fatty amine as described above with acrylonitrile, then hydrogenating to the amine.

Fatty Amido-Amines and Imidazolines. These are manufactured by reacting selected fatty acids with polyamines such as diethylenetriamine or n-amino ethyl piperazine. Polyamines can range from two nitrogens up to 6 or more in some complex polyamines. The polyamine may contain a hydroxy group, which will impart some water solubility.

Oxyalkylated Amines. The amido-amines, primary amines, and imidazolines may be reacted with ethylene oxide to make them water soluble. From one to many mols of ethylene, propylene, or butylene oxide may be added depending upon the degree of water solubility and oil-water partitioning desired. The addition of alkylene oxides to amines forms tertiary amines, which may be quaternized.

Quaternaries. Quaternary ammonium salts are made from tertiary amines such as dimethyl fatty amines or oxylakylated amines, and an organic sulphate or halide. Methyl sulphate or benzyl chloride are commonly used.

Other Amine Derivatives. Amines may be modified by adding phosphorus, sulphur, or other elements to the molecule to improve adsorption to the metal surface. They may be further reacted with epoxides, urethanes, and other crosslinking agents to build molecular weights. The amines are then blended with dimerised fatty acids, etc. to achieve crossbonding of the molecules on the metal surface. They may be blended with low molecular weight inorganic and organic acids to make them water soluble or dispersible. Nitrite and phosphate groups may be added to achieve some surface passivation.

4. FORMULATING FINISHED INHIBITORS

After the proper molecules are selected or synthesised they must be formulated into a usable or "finished" product. The amines are viscous liquids or solids and must be diluted with a proper solvent. The solvent may be an aqueous blend(water, alcohol, or coupling agents) for water solubility. Coupling agents include glycols, glycol ethers, alcohols, esters, and ketones. They are selected to form a stable solution of the amines and other additives, and have the proper degree of water and brine solubility or dispersibility. The inhibitor must also meet requirements for pour point and flash point.

Solvents. A hydrocarbon solvent will be used for an oil soluble or an oil soluble water dispersible inhibitor. Aromatic solvents are preferred since they have much better solvating properties than aliphatic solvents. Pure aromatics such as xylene and toluene are rarely used because of cost, so by-product aromatic solvents from still bottoms, BTX units, from cumene manufacture, and some gas oils and light cycle are used. They may contain some polycyclic materials and anthracenes, so pour point must be considered.

In special instances aliphatics such as diesel will be used.

Other Additives. The formulations may contain special types of surface active agents to aid in dispersing the inhibitor, preventing emulsions, and reducing the time it takes for the amine to "wet" the metal surface. They may also aid in penetrating loose deposits to the metal underneath. These may include sulphonic acids and sulphonates, oxyalkylated materials such as amines, alkyl phenols, fatty acids, and amides, amphoterics, etc.

Attachment 2 lists many of these products.

In addition to the amine there will be materials added to increase film strength such as fatty acids. Low molecular weight organic acids and hydroxy acids are added to achieve water solubility or dispersibility.

Wetting agents may be added to help the amine contact the metal surface and penetrate loose deposits. An emulsifying surfactant/dispersant may be added so that the inhibitor will form a stable dispersion in well fluids. Conversely, a demulsifier may be added to prevent formation of stable oil-water emulsions that can upset production and processing facilities.

Vapor Phase Inhibitors. These include low molecular amines such as ethyl amine, diethyl amine, and triethylamine. They have low boiling points and high vapor pressures and are blended into the corrosion inhibitor formulation for a combination product.

5. MONITORING

The inhibition program must be monitored continuously to be sure that the desired protection is achieved. Iron counts will give a fair indication of the amount of corrosion occurring and actual iron loss can be estimated. It will not tell if pitting is occurring. Manganese analyses can be run to substantiate the iron counts.

Corrosion coupons, LPR, and resistance probes may be installed in the system. In pipelines an intelligent pig can be run at selected intervals to see the actual corrosion and pitting.

Acoustic emission and magnetic flux devices may be used to monitor metal loss and pitting.

6. CONCLUSIONS

In conclusion, the inhibitor must meet certain requirements for each specific application such as stability against temperature, time, and exposure to the corrosive environment. It must function at low concentrations and be easy to apply. Solubility characteristics must be designed for each application, and the inhibitor must be pumpable at the system temperature. It must be compatible with other chemicals in use, and must meet performance specifications- it must inhibit corrosion! It must also be compatible with the system in which it is used, and not cause system upsets. It cannot be too toxic, and the flash point must be within specifications. Raw materials must be readily available and not too expensive, and manufacturing processes capable of control and reproducibility.

It must conform with the application methods available, and monitoring procedures must be straight forward and easily performed.

All these parameters must be the guidance for inhibitor development. All testing and selection of materials must be based strictly on the requirements so that costs and development time are kept to a minimum.

Conformance to requirements is the only valid definition of quality.

ATTACHMENT 1

CHEMICAL RAW MATERIALS FOR INHIBITOR SYNTHESIS

AMMONIA	NH_3
ACRYLONITRILE	$CH_2=CHCN$

AMINES

Ethyleneamines:

Ethylenediamine(EDA)	$NH_2C_2H_4NH_2$
Diethylenetriamine(DETA)	$NH_2C_2H_4NHC_2H_4NH_2$
Triethylenetetramine(TETA)	$NH_2C_2H_4NHC_2H_4NHC_2H_4NH_2$
Tetraethylenepentamine(TEPA)	$H_2C_2H_4NHC_2H_4NHC_2H_4NHC_2H_4NH_2$

High molecular weight condensation products of the above

*N-Aminoethylpiperazine

```
   /CH2-CH2\
HN          NC2H4NH2
   \CH2-CH2/
```

*N-Aminoethylethanolamine $NH_2C_2H_4NHC_2H_4OH$

*Hexamethylene diamine $NH_2C_6H_{12}NH_2$

*Cyclohexamethylene diamine

```
H   /CH2-CH2\   H
N-CH         HC-N
H   \CH2-CH2/   H
```

*Fatty primary amines RNH_2

*Fatty secondary amines

```
R
 \
  NH
 /
R
```

*Fatty tertiary amines

```
R
 \
  NR
 /
R
```

*Fatty diamines $RNHC_3H_6NH_2$

*Oxyalkylated amines

$$H(OC_2H_4)_nN\begin{matrix}/(C_2H_4O)_nH\\ \backslash(C_2H_4O)_nH\end{matrix}$$

Pyridines
*alkyl pyridines from coal tar and steel manufacture
*alkyl pyridine and derivatives from byproducts of other chemicals

Organic Acids
*acetic
*hydroxyacetic
*naphthenic
*oxidized petroleum fractions and waxes

Fatty Acids
*vegetable and animal fatty acids:
C_{12} to C_{16} Coco fatty acids; palmitic, lauric, etc.

*C_{18} acids
oleic- one double bond
linoleic- two double bonds
linolenic- three double bonds

*Dimerized fatty acids

$$\begin{matrix}CH_3(CH_2)_nCH(CH_2)_nCOOH\\ \backslash\\ CH_3(CH_2)_nCH(CH_2)_nCOOH\end{matrix}$$

*Rosin acids
abietic
neo-abietic

Alpha Olefins
*C_{12}^{+} alpha olefins

Acetylenic Alcohols
propargyl alcohol
ethyl octynol

<u>Reactions</u>

*Manufacture of ethyleneamines:
react ethylene dichloride with ammonia

(amine hydrochloride)

$$ClC_2H_4Cl + 2NH_3 \text{ -----> } NH_2C_2H_4NH_2{:}2HCL$$

$$NH_2C_2H_4NH_2{:}2HCL + Ca(OH)_2 \text{ ----> } NH_2C_2H_4NH_2 + CaCl_2$$

*Amidization:
react ethylenamines with fatty acids to form amido-amines

$$NH_2C_2H_4NH_2 + RCOOH \text{ ----> } NH_2C_2H_4NHOCR$$

*Cyclization:
heat amido-amines, drive off water of reaction, form imidazolines

```
NH2C2H4NHOCR ---->CH2-----N
                   |      ||   + H2O
                  CH2    CH2
                    \    /
                      N
                      R
```

*Oxyalkylation:
addition of ethylene and propylene oxide to amines, fatty acids, amides, and to phosphoric acid

```
RNH2 +H2C---CH2 -----> RNHCH2CH2OH
        \ /
         O
```

*Quaternization:
react tertiary amines with halides or sulphates to form quaternary ammonium salts

```
                            -             -
                          |  R            |+
                          |  |            |
R3N + C6H5CH2Cl ---> R    | -N- C6H5CH2   |Cl-
                          |  |            |
                          |  R            |
                            -             -
```

*Epoxidation:
form fatty epoxides from alpha olefins for further reaction

*Amine oxide formation:
react tertiary amines with hydrogen to form amine oxides

*Dimerization:
polymerization of fatty acids containing double bonds by reacting with ozone to form dimer and trimer carboxylic acids

*Addition of other reactive groups:
phosphorus compounds for passivation properties
sulphurization of fatty acids for film strength
nitrite-nitrate for passivation and oxygen resistance

ATTACHMENT 2

ADDITIVES FOR FORMULATING CORROSION INHIBITORS

SURFACTANTS FOR WETTING AND DISPERSING

Anionics:
*Sulphonated fatty acids
*Sulphonated petroleum fractions
*Alkyl aryl sulphonates
*sulphated fatty alcohols
*sulphated fatty ethers
*sulphosuccinic acid derivatives
*Phosphate esters of fatty alcohols and their ethoxylates
*Fatty taurates

Cationics and amphoterics:
*Quaternaries
*Oxyalkylated amines(low ethylene oxide content)
*Sulphobetaines

Nonionics:
*Oxyalkylated alkyl phenols
*Oxyalkylated fatty acids and alcohols
*Oxyalkylated amines(high ethylene oxide content)
*Fatty triethanolamides and their ethoxylates
*Fatty acid glycol esters
*Sorbitol-fatty acid esters
*Oxyalkylated sorbitol-fatty acid esters
*Ethylene oxide/propylene oxide block polymers

Surfactants for Emulsion Prevention

*Propylene/ethylene oxide adducts of glycols, amines, and resins
*Sulphonates
*Polyglycol esters, urethanes, and epoxides

Coupling Agents for Stability, Solubility, and Pour Point

*Alcohols- isopropanol, octanol, etc.
*Glycols- ethylene, diethylene, etc.
*Glycol ethers

*Esters
*Ketones

Miscellaneous

*Phosphonates and polyacrylates for scale control/dispersion
*Oxygen scavengers

BIBLIOGRAPHY

1. Ir. M. J. J. Simon Thomas, C. de Waard Ph. D. L. M. Smith MA. Ph. D. MIM C. Eng. "Controlling Factors in the Rate of Corrosion" Corrosion UK, Brighton England 26-28 October, 1987

2. "CO_2 Corrosion in Oil and Gas Production- Selected Papers, Abstracts, and References" National Association of Corrosion Engineers, 1984

3. "Advances in CO_2 Corrosion, Volume 1," National Association of Corrosion Engineers, 1984

4. "Advances in CO_2 Corrosion, Volume 2", National Association of Corrosion Engineers, 1985

5. "Corrosion Inhibitors", National Association of Corrosion Engineers, 1973

6. "Corrosion Control in Petroleum Production, TPC-5", National Association of Corrosion Engineers, 1979

7. C. de Waard, D. E. Milliams, "Prediction of Carbonic Acid Corrosion in Natural Gas Pipelines" First International Conference on the Internal and External Protection of Pipes, BHRA Fluid Engineering, University of Durham, Sept. 1975

8. A.K. Dunlop, H.L. Hassell, P.R. Rhodes, "Fundamental Considerations in Sweet Gas Well Corrosion", Corrosion/83, Anaheim Ca. National Association of Corrosion Engineers, April 1983

9. Elisabeth Eriksrud, Terje Sontvedt, "Effect of Flow on CO_2 Corrosion Rates in Real and Synthetic Formation Waters", Corrosion/83, Anaheim Ca. National Association of Corrosion Engineers, April 1983

10. "Corrosion Engineer's Reference Book", National Association of Corrosion Engineers, 1980

11. A. W. Peabody,"Control of Pipeline Corrosion" 1976

12. C. C. Patton,"Oilfield Water Systems,"1981

13. L. W. Gatlin, H. W. Endean, "Water Distribution and Corrosion in Wet Gas Transmission Systems", Corrosion/75, National Association of Corrosion Engineers

14. "Forms of Corrosion Recogniton and Prevention" NACE Handbook 1, National Association of Corrosion Engineers 1982

15. "Modern Electrical Methods for Determining Corrosion Rates", NACE Publication 3D170,National Association of Corrosion Engineers

16. "Carbonic Acid Corrosion of Steel", Corrosion 1975, Vol 31 (no. 5)

17. James E. Donham,"Recent Developments in Corrosion Inhibitors and Their Use", Canadian Western Region Conference, Edmonton, Alberta, 1985

18. S. Evans, J.M. Phelan, M.E. Williams, "Batch Treatment of Offshore Wells in the East Cameron and Vermilon Area", 17th Annual Offshore Technological Conference, Houston, Tx. May 1985

19. T. Murata, E. Sato, R. Matsuhashi, "Factors Controlling Corrosion of Steels in CO_2 Saturated Environments", Corrosion/83, Anaheim, Ca. National Association of Corrosion Engineers, April 1983

20. "Petroleum Engineers Handbook", Society of Petroleum Engineers, 1987

21. D. F. Story , "Recent Advances in On-Line Inspection and the Related Benefits to Pipeline Operators", Second Annual Seminar, Pipeline Protection and Maintenance Seminar, October 1987, Bergen, Norway. A/S Norwegian Applied Technology

22. Leonard C. Choate, "In Situ Cleaning and Coating of Offshore Gas Pipelines", Second Annual Seminar, Pipeline Protection and Maintenance Seminar, October 1987, Bergen Norway. A/S Norwegian Applied Technology

23. B. C. Goff, G. L. Higgins, "Recent Developments in `In-Situ' Internal Cleaning and Coating of Pipelines", Second Annual Seminar, Pipeline Protection and Maintenance Seminar, October 1987, Bergen, Norway. A/S Norwegian Applied Technology

24. G. Schmitt, "Fundamental Aspects of CO_2 Corrosion", Corrosion/83, Anaheim, Ca. National Association of Corrosion Engineers

25. J.L. Crolet, M.R. Bonis, "The Role of Acetate Ions in the CO_2 Corrosion", Corrosion/83, Anaheim, Ca. National Association of Corrosion Engineers

26. Peter Griffin, "Multiphase Flow in Pipes" Journal of Petroleum Technology, March 1984

27. Milloud El-Arabi, Adam Mazur, "Flowline Failure Resulting from Erosion-Corrosion in a High Production Oil Well", Materials Performance, September 1985

28. James P. Brill, "Multiphase Flow in Wells", Journal of Petroleum Technology, January 1987

29. John Shaw, "Corrosion Inhibitor Treatments", First Annual Seminar, Pipeline Protection and Maintenance, A/S Norwegian Applied Technology, 8. October 1986.

30. K. Minami, J.P. Brill,"Liquid Holdup in Wet-Gas Piplines",SPE Production Engineering, February 1987

31. Thomas E. Chamberlain," Practical Application of Potentiodynamic Polarization Curves in Oil Well Corrosion", Corrosion/85, Boston, Mass. National Association of Corrosion Engineers

32. J.A. Martin, F.W. Valone, W.C. Haltmar,"The Determination of Corrosion Inhibitor Residual Roncentration", Materials Performance, May 1986

33. H.J. Choi, R.L. Cepulis, "Inhibitor Film Persistency Measurement by Electrochemical Techniques in Laboratory-Simulated Environments", International Symposium on Oilfield and Geothermal Chemistry, Phoenix, Ar. April 1985. Society of Petroleum Engineers

34. J.L. Crolet, M.R. Bonis, "pH Measurements in Aqueous CO_2 Solutions under High Pressure and Temperature", Corrosion, February 1983

35. "Recommended Practice: Preparation and Installation of Corrosion Coupons and Intepretation of Test Data in Oil Production Practice",National Association of Corrosion Engineers Standard RP-07-75

36. "Test Method: Dynamic Corrosion Testing of Metals in High Temperature Water", National Association of Corrosion Engineers Standard TM-02-74

37. "Test Method: Autoclave Testing of Metals in High Temperature Water", National Association of Corrosion Engineers Standard TM-01-71

38. "Recommended Practice: Control of Internal Corrosion in Steel Pipelines and Piping Systems", National Association of Corrosion Engineers Standard RP-01-75

39. Marcel Decker, Inc. "Fatty Acids and their Industrial applications"

40. "Committee Report: Wheel Test Methods for Evaluating Film Persistent Inhibitors for Oilfield Applications", National Association of Corrosion Engineers Task Group T-1D-8 Technical Committee Report

42. "Committee Report: Proposed Standardized Laboratory Procedure for Screening Corrosion Inhibitors for use in Oil & Gas Wells", National Association of Corrosion Engineers Publication 55-2

43. J.E. Donham, US Patent 3,699,118, "Reaction Products of Substituted Imidazolines and Amino Tri(lower alkylidene-phosphonic acids)"; Assigned to Amoco Production Company

44. "The Thesaurus of Chemical Products, Volumes I and II", Chemical Publishing Company, Inc. 1986

Corrosion Inhibitor Testing: Principles and Practices

By A. D. Mercer

DIVISION OF MATERIALS APPLICATIONS, NATIONAL PHYSICAL LABORATORY, TEDDINGTON, MIDDLESEX TW11 0LW, UK

1 INTRODUCTION

The recent ISO definition of an inhibitor is "Chemical substance which decreases the corrosion rate when present in the corrosion system at a suitable concentration without significantly changing the concentration of any other corrosive agent".[1] This definition thus excludes chemicals which can reduce the corrosion rate by affecting the composition of the environment, e.g., oxygen scavengers, chemicals that alter the pH and chemicals that change the hardness or scale forming properties.

Thus, considerations of the principles and practices of corrosion inhibitor testing will be restricted to those chemicals which meet this definition. A feature of these chemicals is that the concentrations required for inhibition cannot be calculated on the basis of known chemical laws or equilibria. Such calculations can be made for the concentrations of oxygen scavengers, pH adjustments, etc. Since the concentrations required cannot be calculated it is necessary to arrive at recommended values on the basis of tests and experience. It is the purpose of this paper to consider the principles and practices that are involved in assessing the effectiveness of inhibitors.

Inhibitors and Inhibited Products

At this point it is necessary to distinguish between inhibitors per se, that is, clearly identifiable chemical compounds singly or in various combinations on the one hand, and products containing inhibitors on the other hand when the chemical identity of the inhibitor may not be disclosed and will often be present with other chemicals having other functions.

An example may be taken to illustrate this point which is of some importance. Benzotriazole is widely used as an inhibitor of the corrosion of copper in aqueous systems and various tests can be devised to determine the way in which its effectiveness is governed by environmental conditions such as temperature, movement, salinity etc. This chemical may then be included in formulations with other additives for industrial applications in cooling systems, hydraulic fluids, wrapping papers, etc. and tests must be devised that are appropriate to these particular conditions of use.

These "service" tests are then found to be of two types. The first type consists of those tests which are developed by users and customers of inhibited products on a quality assurance basis: these tests will often be employed only by the organisation using the product but may nevertheless be reported in the scientific and technical literature. The second type consists of those tests which have been developed - and often on the basis of those of the first type - into industry standards and also into national and international standards.

The above sequence of events can be summarised as a sequence of inhibitor development: laboratory testing of a single chemical or formulation of chemicals → company standards and tests for products containing those chemicals → institute or association standards and tests, e.g. Institute of Petroleum (IP), American Society for Testing and Materials (ASTM), National Association of Corrosion Engineers (NACE), etc. → tests described in national and international standards e.g., BSI, DIN, ISO etc.

All test procedures, except those of a simple pass or fail criterion based on visual perception, must be based on measurement and consideration of this aspect is particularly important in corrosion and its inhibition. This is because there are a number of criteria that can be used. These include the measurements of mass loss, corrosion currents, pit depth, time to failure of mechanically loaded components, changes in metal content of the environment and others. Each of these measurement methods attracts its own units and the selection will depend on the purpose for which the inhibitor is being used.

2 TEST PARAMETERS

In the development of test methods for inhibitors it is important to consider the parameters necessary for the characterisation of inhibitors. This is particularly important in laboratory testing of single chemicals in isolation before the situation is complicated by the presence of other chemicals in a marketed product. Furthermore, laboratory testing with a range of parameters should reveal those to which an inhibitor is particularly sensitive and which therefore need particular attention in the development of standard tests.

Nature of Metal

This is of particular importance because, with few exceptions, inhibitors are highly specific in their action towards metals. While chromates are universal in their action the same is not true of nitrites which protect ferrous metals but not zinc or aluminium. Test procedures should therefore always take into account all metals and alloys with which the inhibitor might come into contact. In this matter care should also be taken in the selection of alloy composition as variations in this can affect inhibitor performance. In neutral solutions, for example, it is found that sodium benzoate will protect mild steel but not cast iron whereas the related compound sodium cinnamate will protect both materials.[2] The existence of bimetallic contacts must also be considered. These may arise deliberately or inadvertently as where mild steel is exposed at the edges of cut galvanised strip. Chemicals which in other circumstances may protect steel can fail to do so when it is in contact with zinc.

Surface Condition of the Metal

This is also of importance in testing inhibitors. For clean, bright surfaces it may be found that the concentration of inhibitor required to maintain this condition is greater for, say, a shot blasted surface than a finely abraded surface.[3] Of equal, if not greater, importance will be the presence of corrosion products as will often be the case in a real system. The existence of such products will be an indication that some part of the system is actively corroding. Steps must then be taken to test inhibitors in these conditions since it cannot be assumed that inhibitors that function satisfactorily with clean surfaces will necessarily be effective with rusting surfaces. Some laboratory tests to demonstrate this point with the example of borax as an inhibitor for corroding zinc have been reported.[4]

Dimensions and Geometries

These are points requiring attention in any test procedure. A related matter is that of specimen area to solution volume because the latter will control the total amount of inhibitor that will be available. The dimensions may also be dictated by the stock material available: a test procedure specifying abraded sheet specimens is difficult to achieve if only cast bars are available. The test specimens, or coupons, can in fact take various forms: sheet, wire, bar, tubes etc but in each case attention has to be given to the method of mounting particularly when electrical connections need to be made for electrochemical measurements.

Test Apparatus

The form of the test apparatus can be selected from a wide spectrum. In the simplest case test specimens can be immersed in test tubes, beakers or other convenient small containers and much useful work is conducted in this way. Effects of temperature can be studied by placing such containers in water baths at controlled temperatures. Caution must be exercised with the use of borosilicate glass at temperatures above about 60°C because of the distinct possibility of silicates being leached into the test medium. Silicates have inhibiting properties and could provide misleading results. Quartz containers are more suitable in these conditions as indeed are some polymeric materials such as ptfe. The effects of glass dissolution on the corrosion of mild steel in non-inhibiting conditions has been described.[5]

Test Duration

The length of the test should relate to the conditions in which the inhibitor is to be used. If required for metal pickling processes the test period need be no longer than minutes, for boiler cleaning processes perhaps hours and for engine cooling systems months will be more appropriate.

Composition of the Test Medium

This factor should also be considered in relation

to the type of application for which the inhibitor is being considered. If an inhibitor is intended for use in a refrigeration brine then clearly testing should be conducted in this medium. In the case of more dilute solutions of salts, as mains water might be, it is important to characterise the inhibitor performance over a range of concentrations of the appropriate salts since such a range will exist in practice. For example, natural waters will have chloride levels from as low as 10 ppm to as high as 200 - 300 ppm or even higher outside the UK. The effect of pH and variations in this parameter should also be taken into account.

Persistence of Inhibition

This is of importance in once-through systems in which the maintenance of a fixed concentration of inhibitor may be difficult to achieve. Tests must then be devised in which this feature of inhibitor performance is included.

Inhibitor effectiveness may also be influenced by movement of the environment and in fact this may enhance protection up to a certain point. However, high rates of flow may lead to the occurrence of erosion and even cavitation effects and these must be considered in appropriate cases.

3 PUBLISHED TEST METHODS

As indicated above these may be broadly divided into those that have gained acceptance to the point where they are incorporated into national or company standards and others which although adequately described in the literature have not received this type of recognition. The latter may be referred to as non-standard tests and some are described below before a discussion of standard tests.

Non-standard Tests

Neutral Solutions. A simple procedure involves exposing test specimens to a fixed volume of the inhibited solution and comparing the corrosion behaviour with that of a similar specimen in the solution without an inhibitor. A test used in the author's laboratory uses, for example, 30 x 25 x 1.6 mm abraded sheet specimens of the required metal resting at an angle of about 60° in 60 cm^3 of solution in glass test tubes. Variations of this arrangement have been described but in all cases the extent of inhibition is assessed from changes in mass of the specimen and from observations.[6] Other types of test rely solely on observations and are particularly valuable for screening tests where large numbers of potential inhibitors need to be assessed. Hersch et al. used steel discs placed in glass dishes and covered with inhibitor solution.[7] In another series of tests these authors used 0.03 cm^3 drops of inhibitor solution of various concentrations placed on a degreased steel plate. The critical concentration of inhibitor was the lowest at which no corrosion took place under each of triplicate drops.

This general approach to inhibitor testing is used by one supplier of inhibitors for domestic central heating systems. Small volumes of solution are drawn off from the system into a jam jar containing a clean nail. If protection is maintained for a 24 hour period the test is repeated with the sample diluted by a measured amount and so on until rusting occurs. This provides a measure of the inhibitor reserve in the system.

These tests are conducted in essentially static conditions. Forced movement can be obtained by rotating the test specimen or by keeping the specimen in a fixed position and moving the liquid by stirring, gas bubbling or pumped circulation.

Thus, a number of rigs have been described in which the technique has varied from flat specimens in glass vessels, to small test-heat exchangers,[8,9] simulated cooling towers and semi-scale plant.[10,11] Heat transfer conditions have been provided by mounting tubular specimens in an annular tube or shell through which a hot fluid is passed. An alternative method is to insert heated metal plugs in a flowing inhibited solution.[6,12]

Acid Conditions. In acid solutions similar techniques can be used and one procedure has been described in which tests in inhibited hydrochloric acid are carried out at 121°C with flow velocities up to 44 ms^{-1}.[13]

A circulating corrosion loop for testing inhibitors for acid cleaning processes has been described by Trabanelli et al.[14] The rig has a filling vessel, two centrifugal pumps in parallel and three parallel branches for testing tubular specimens using mass-loss and electrochemical methods. The total volume of the system is 135 litres, maximum acid temperature 75°C and velocity 4 ms^{-1}. This rig can be used with hydrofluoric acid since all wetted parts are coated with ptfe.

Oil and Gas Production and Transport. Various procedures have been described for the selection of inhibitors for crude oil pipelines. A 'bubble test' uses two vertical and parallel polished specimens of a pipeline steel mounted in a 1 l flanged beaker containing a 20:80 oil/water mixture.[15] The system is thoroughly deaerated with hydrogen and then CO_2 is introduced to bring the pH to the required value. The inhibitor is introduced after a return to hydrogen purging and various electrochemical procedures are used to assess the performance of the inhibitor in the separated water phase. Another system takes the form of a compact low volume (3 l) test loop connected to two reservoirs containing respectively the brine and oil.[15] Specimens are of tubular form and inserted into the pipework of the loop.

The wheel test method is a dynamic test in which volumes of about 200 cm^3 volume containing weighed test specimens and a two-phase fluid saturated with the appropriate gas (CO_2 or H_2S) are loaded on to a wheel that rotates inside a temperature controlled chamber.[16] In some cases the specimens are transferred from an inhibited to a non-inhibited fluid so that the persistence of inhibition can be assessed.

Volatile Corrosion Inhibitors. Volatile corrosion inhibitors (VCI) have been evaluated by the methods described in 1951 by Wachter et al. in the USA,[17] and by Stroud et al.,[18] in the UK. In the former, air of required humidity was drawn through tubes lined with paper impregnated with VCI and then passed over the metal test specimens. In the UK test specimens were suspended above aqueous solutions of VCIs at appropriate temperatures with the option of SO_2 injection into the vapour phase to produce more aggressive conditions. Other tests for VCI's have been described in the USSR.[19]

Standard Tests

The term standard tests is used to describe those that can be identified as recognised procedures from national or other bodies. Tests of this nature are often designed to determine the corrosion behaviour of a product or system but may be included as inhibitor test methods since the protection results from the use of chemicals recognised as inhibitors.

Acid Conditions. A simple test for inhibited hydrochloric acid is described in a UK specification.[20] In this a degreased and weighed mild steel specimen is immersed for 6 hours in a 1:4 acid:water mixture at 25°C with the mass change of the specimen after this period used a a measure of the inhibition.

Engine Coolants. Much of the work on standard test procedures for inhibited aqueous solutions has been associated with the subject of engine coolants. Such testing is of the approval type, i.e., rather than the research type for which various other tests are used, in which inhibited coolants are assessed for their behaviour towards the metals and alloys found in engine cooling systems. Tests are essentially of two types: glassware tests using static - but aerated - solutions in 1 l vessels containing specimens of the relevant metals, and recirculating rig tests.[21,22,23,24] In these rig tests a coolant is cirulated in a flow loop consisting of a reservoir, pump and connecting hoses in which test specimens are placed. The ASTM method uses actual engine components,[22] e.g. radiators, in the rig while the BSI method uses components specially designed for the rig.[23] Problems of reproducibility and repeatability continue to be associated with complex rig tests and this is perhaps not surprising in view of the operating conditions which are often difficult to control Nevertheless, these rigs have been extensively used and improvements are always being considered.

Special tests for cavitation resistance had been developed in the FRG,[25] and in the USA.[26]

Soluble Oil Emulsions, Hydraulic Fluid, Mineral Oils, Fuels. Procedures to determine the corrosion inhibiting properties of soluble oil emulsions are given in the so-called chip-filter paper method of IP 287/83 and DIN 51 360 (Part 2).[27,28] Cast iron chips are placed on filter paper wetted with the emulsion and the area of rust stained paper is assessed after 2 h. This is repeated with various dilutions until a significant increase in stained area results: this is referred to as the break point. The composition of the water for preparing the emulsion is specified.

The inhibitive properties of water based hydraulic fluids can be assessed using the procedure of IP 329/84.[29] Weighed test specimens of steel, copper, brass, aluminium alloy, cast iron, zinc and cadmium are assembled in a manner similar to that used in engine coolant testing. However, for hydraulic fluids partly immersed conditions are used, the test is at 50°C and the duration is 330 hours.

A standard test method for the rust preventing characteristics of inhibited mineral oil in the presence of water is described in ASTM D665-83 and the corresponding IP 135/64(82) as well as BS 2000:Part 135:1983.[30,31,32] In this a mixture of 0.3 l of oil and 0.03 l of distilled or artificial sea water is stirred at 60°C in a beaker containing a polished steel electrode which is fully immersed: the test period is usually 24 hours and visual inspection is made of the specimen at the end of this period.

A US military specification Mil-I-25017-C of 1971 for fuel soluble corrosion inhibitors uses a modification of the D 665 test. It uses hot rolled steel spindles immersed in an agitated mixture of hard water and iso-octane at 38°C for 5 hours. Test methods for oil soluble inhibitors for fuels are also described by EMPA and FTM procedures [33,34] The former uses round, abraded steel test specimens potitioned horizontally in glass beakers containing 200 cm^3 of inhibited oil that has been thoroughly shaken for 5 min with 1 cm^3 of 0.5% sodium chloride in deaerated distilled water. The test is of 12 weeks duration. The FTM list uses triangular hollowed steel test specimens which rest on the bottom of a glass vessel holding 0.5 l of sodium chloride in deaerated distilled water. The water phase is allowed to settle out and is removed. The oil is placed in the test vessel and after 1 hour 200 cm^3 of the water phase is put into each of the hollows in the specimens which are then removed for inspection at two week intervals over a period of 12 weeks.

CONCLUSION

This review has indicated some principles that need to be considered in developing tests for corrosion inhibitors. Tests are of three distinct types: laboratory tests for characterising inhibitors, published tests used by companies, associations, etc. for evaluating inhibited products and national and international standards.

REFERENCES

1. Corrosion metals and alloys - Terms and definitions ISO 8044- 1986
2. A. D. Mercer, 5th European Symposium on Corrosion Inhibitors, University of Ferrara, Ferrara, Italy, September 1980, Proceedings 5 SEIC, Annali Univ., Ferrara, NS, Sez V, Suppl. (7), 1980, 563-581.

3. D. M. Brasher and A. D. Mercer, Br. Corros. J., 1968, 3, 120-129.
4. J. G. N. Thomas and A. D. Mercer, Materials Chemistry and Physics, 1984, (10), 1-20.
5. A. D. Mercer and G. M. Brook, La Tribune de Cebedeau, 1978, 31, (417-418), 299-306.
6. G. Butler, A. D. Mercer and G. M. Warren, 6th European Congress on Metallic Corrosion, London, Sept 1977, Soc. Chem. Ind., London, 1977, p 349-356.
7. P. Hersch, J. B. Hare, A. Robertson and S. A. Sutherland, J. Appl. Chem., 1961, 11, 246-271.
8. A. J. Freeman, W. J. Ryzner and J. D. Knapp, Mater. Prot., 1962, 1, (10), 22-30.
9. D. C. Zecher, Mater. Perform. 1976, 15, (4), 33-37.
10. R. W. Lane, Mater. Perform, 1973, 12, (9), 27-30.
11. P. Sury and R. Hiltbrunner, Sulzer. Tech. Rev., 1978, (4), 172-178.
12. A. D. Mercer in "Engine coolant testing: state of the art", STP 705 (ed W. H. Ailor), American Society for Testing and Materials, Philadelphia, PA, 1980, 51-80.
13. W. E. Billings, J. A. Knox and D. Morris, Mater. Prot., 1963, 2, (8), 58-62.
14. G. Trabanelli, F. Zucchi, A. Frignani, M. Zucchini and G. Brunoro, 5th European Symposium on Corrosion Inhibitors, University of Ferrara, Ferrara, Italy, September 1980, Proceedings 5 SEIC, Annali Univ., Ferrara, NS, Sez V, Suppl (7), 1980, 1141.
15. J. Bown and D. Harrop in Proc. Conf. on "The development and use of corrosion inhibitors", OYEZ Scientific and Technical Services Ltd, London, Dec, 1983.
16. Mater. Perform, 1982, 21, (12), 45-47, NACE publication ID 182, National Association of Corrosion Engineers, Houston, Tex.
17. A. Wachter, T. Skei and N. Stillman, Corrosion, 1951, 7, 284-294
18. E. G. Stroud and W. H. J. Vernon, J. Appl. Chem.,1952, 2, 178-184.
19. S. Z. Levin, S. A. Gintzberg, I. S. M. Dinner and V. N. Kuchinsky, Ann. Univ, Ferrar (Nuove Serie), Section 5, 1966, Suppl 4, (Proc. 2nd European Symp. on Corrosion Inhibitors (SEIC), Ferrara, 1965), 765-776.
20. Defence Standard 68-27/Issue 2, "Scale removing compounds (for hot water installations)", Ministry of Defence, Directorate of Standardisation, Stan 1, Glasgow, 1977.
21. BS 5117, "Testing corrosion inhibiting, engine coolant concentrates ('antifreeze'), Section 2.2, 1985, "Glassware test".
22. ASTM D 2570, "Simulated service corrosion testing of engine coolants".

23. BS 5177, "Testing corrosion inhibiting, engine coolant concentrates ('antifreeze'), Section 2.3, 1985, "Recirculating rig test".
24. R. Schulmeister and H. Speckhardt in "Engine coolant testing: state of the art". STP 705, (ed. W. H. Allor), American Society for Testing and Materials, Philadelphia, Pa, 1980, 81-100.
25. G. A. Liebold and K. W. Sterke, in "Engine coolant testing: second symposium". STP 887, (ed. R. E. Beal), American Society for Testing and Materials, Philadelphia, Pa, 1986, 176-188.
26. ASTM D 2966, "Test for cavitation erosion-corrosion characteristics of aluminium in engine coolants using ultrasonic energy".
27. IP 287/83, "Rust prevention characteristics of soluble oil emulsion chip filter paper method".
28. DIN 51 360, Part 2, "Testing of cooling lubricants, determination of corrosion preventing characteristics of water-mixed cooling lubricants: chip/filter paper method".
29. IP 329/84, "Corrosive properties of water-based hydraulic fluids"
30. ASTM D 665-83, "Standard test method for rust preventing characteristics of inhibited mineral oil in the presence of water".
31. IP 135/64(82), "Rust-preventing characteristics of steam turbine oil in the presence of water" (and ASTM D 655-82).
32. British Standard Methods of Testing for Petroleum and its Products: BS 2000, Part 135:1983, "Rust-preventing characteristics of steam-turbine oil in the presence of water".

Factors Influencing the Formulation of Treatments for Use in Heating and Cooling Systems

By R. D. Davies

TECHNICAL SERVICES DEPARTMENT, FERNOX MANUFACTURING COMPANY LIMITED, BRITANNICA WORKS, CLAVERING, ESSEX CB11 4QZ, UK

1 INTRODUCTION

There are many factors influencing the formulation of water treatments for use in mixed metal systems mainly because of the emerging demands for "total" water treatment in the light of the changing design of heating and cooling plant. The employment of smaller plant with ever improving efficiencies, the use of different alloys and new combinations of mixed metal components together with the increased use of non-metallics, all create new corrosion and water conditioning problems. The increased awareness of the ecological factor and the pressures of economics are also important influences. As a result, all too often, traditional water treatment formulations are proving to be inadequate. These new demands must create changes and increased performance of formulation design.

Traditionally designers of heating and cooling plant did not think that chemical treatment of system waters was necessary as plant was constructed from components of such wall thickness that any corrosion occurring would be of minimal importance. It was assumed that the formation of sludge within the system would not affect the working of the pumps and valves as they were designed with large apertures and tolerances. It was in the light of blissful ignorance that much plant was built to be self destructive.

The first problems encountered were the failure of steam raising boilers resulting in death and injury. The use of potatoes and natural organic tannins as sludge dispersants became commonplace soon to be followed by the additional use of sodium phosphates augmented by the periodic addition of soda ash. Whilst the addition of these chemicals did much to reduce the problems, the lack of understanding and chemical control led to the effects of caustic embrittlement and resultant explosions. Problems in this area were aggravated by the development of larger boilers with higher working pressures and heat transfer rates resulting in the formation of new more complex precipitates on the heat exchanger surfaces. It was at that time that the treatment of the boiler feed waters was first contemplated and methods of partial softening and deaeration were explored. The use of sodium sulphite for oxygen removal in the boiler was then recognised as was the need for antifoams which were of organic origins. The total solids content of the boiler was recognised as an important factor and the specific conductance method of measurement became generally employed.

It was approximately at that state of knowledge that primitive treatment of cooling waters was attempted. This was restricted to the use of pH adjustment by the addition of sulphuric acid and alkalies in an attempt to control the scaling in the systems. In answer to some of the problems being encountered in these systems the threshold effects of hexametaphosphates were investigated and the need for biological control was identified and chlorinated phenols were tested.

Simultaneously further discoveries were being made in water treatments for steam boilers. The increased control of silicate deposition using activated magnesium oxide, the use of catalysed sulphites for the rapid removal of dissolved oxygen, together with the use of volatile film forming amines for condensate line protection were all significant advances. Eventually hydrazine was utilised as an oxygen scavanger replacement for sulphites.

As a result of laboratory studies on scale formation, certain organic synthetics were found to stimulate hardness salt crystal distortion. Within a few years low molecular weight acrylic polymers were being developed to replace the more traditional natural tannins and lignins. The eventual development of polyacrylates, polymethacrylates and carboxymethyl cellulose for use as dispersants formed the basis of many of today's weapons for fighting scale and sludge deposition.

The development of chelating organics such as Ethylene-diamine tetra acetic acid (EDTA), Nitrilo triacetic acid (NTA) and diethylene triamine tetra acetic acid (DTPA) was also a significant step in water treatments for both heating and cooling systems. Their ability to prevent and redissolve the scale deposits led to their widespread use in water treatments. Sequestrants were, however, expensive to use, often difficult to control reserves and their evolution to their phosphonate derivatives paved the way for viable threshold treatments with good thermal stability.

As a result of many years of laboratory and field corrosion studies and the development of electrochemical techniques, considerable experience and understanding has been acquired on the use of ions as inhibitors for metals.

2 CORROSION IN HEATING AND COOLING SYSTEMS

The most common type of corrosion is oxidic corrosion and is prevalent in open cooling systems and to some extent in open heating systems due to oxygen diffusion from the feed and expansion tank. Oxidic corrosion has even been observed in closed pressurised heating and cooling systems either due to diffusion through incorrectly sized expansion vessels, or due to the suction of air through microleaks on joints close to the negative pressure side of the pump. The increased use of plastics in underfloor heating systems has also increased the likelihood of oxygen diffusion into the systems.

Electrolytic corrosion occurs in all systems where metallic components are used. This is also evident in "one" alloy metal systems due to the presence of relatively cathodic and anodic crystal areas in the metal surface, or due to the presence of cathodic impurities in the surface of the metal. The presence of dissimilar cathodic metal particles on the metal surface, the stresses built into the worked or cast metal component, the heat flux conditions operating through the metal and the electroless plating of metal from salts present in the system water all provide additional impetus to the electrolytic corrosion process.

Other conditions which promote electrolytic corrosion include the presence of carburised steel and other metals and carbon deposits from flux or drawing lubricants which have been charred during welding, brazing or soldering. These can produce extremely corrosive conditions which can result in rapid component failure.

Galvanic corrosion exists between dissimilar metal components present in the system. The extent and effect of galvanic corrosion is also dependent upon system temperature and many instances have been noted where polarities between dissimilar metal components change and are even reversed with the temperature cycling of the installation.

We have identified five other factors which create additional corrosion problems:-

1. Acidic impurities in the system feed water such as peat acids in some areas. These impurities can be aggravated by residues of descaling acids or by the incorrect use of acids in an attempt to control the pH, or from carry over from incorrectly operating demineralising plant. In evaporative cooling systems, acid impurities from the air can be accumulated due to the scrubbing action of the cascading process.

2. The presence of chlorides and sulphates in the system feed water can promote pitting and crevice corrosion. Chlorides can also originate from malfunctioning water softening plant as well as from soldering and brazing flux residues. Chlorides and sulphates can be introduced into the system as residues from previous acid cleansing regimes or from acids used in pH control. At relatively low levels the anions are of minor importance except where the supply water is soft or has been artificially softened. In these cases rapid pitting corrosion can be promoted. In evaporative cooling and humdification systems, these anions can be concentrated by many times their feed water levels with the resultant tendency for corrosion and selective metal leaching from alloys to occur.

3. Sediments, either from airborne pollutants, or from the precipitation of suspended hardness salts, or from constructional debris, all can create the conditions for crevice corrosion. The formation of such deposits also encourages the action of bacterial attack on the system especially in the form of anaerobic Sulphide Generating Bacteria.

4. The erosion of surfaces due to the system water velocities and in some instances cavitation effects which are aggravated by the presence of suspended solids.

5. The effects of stray electric currents, static charges caused by turbulence and the presence of induced magnetic fields .

All the above corrosion effects can be aggravated by poor system design and installation. The use of water treatments in such systems is in many cases attempting to shut the stable door after the horse has bolted. Typical design faults are the use of galvanised pipework and calorifiers in domestic hot water and heating systems, the use of copper pipework in the presence of aluminium and galvanised metal surfaces, the use of flat bottomed tanks in heating and cooling installations, the installation of small bore pipework under floors with slow water velocity, the use of tight angled bends in areas of high water velocity and many other equally avoidable weaknesses. Similarly the design of both metallic and non-metallic components create highly stressed areas, and many other equally avoidable weaknesses. The siting of evaporative cooling towers and feed and expansion tanks on roofs often near chimneys and other outlets for exhaust gasses can also lead to the formation of system problems outlined earlier.

The effects of these corrosion processes can lead not only to primary corrosion damage such as the perforation of metal components but also to the formation of corrosion products, scale suspensions and precipitates which can cause secondary damage such as the premature failure of pumps, air separators, and solenoid valves. The build up of scale on heat exchangers under high heat flux can lead to the buckling and failure of boilers, or loss of efficiency of the installation. The formation of such precipitates in both heating and cooling systems can also provide excellent micro-environments for the growth of fungi, yeasts and bacteria which can have a profound effect upon the surrounding environment. The costs of corrosion through premature replacement of components, labour and administrative costs and lost plant utilisation contribute an annual bill to the country of many billions of pounds.

The increased costs of energy, water, plant and maintenance means that heating and cooling systems are now being designed for even greater efficiency. In addition installations are more compact and use components of smaller engineering tolerances and tighter performance specs. This has led to a greater need for effective water treatments capable of maintaining trouble free operation and, as a result, many of the traditional approaches to water treatment have had to be adapted or entirely changed.

3 IDEAL PROPERTIES OF A WATER TREATMENT

A water treatment should tolerate and be effective under the following conditions:-

All types of potable water compositions. In most cases this will include various levels of chlorides, sulphates, dissolved gasses, peat acids, hardness salts and cuprosolvency.

Additions of corrosive substances such as soldering and brazing flux residues and sometimes dissolved gasses.

The total absence or liberal presence of oxygen.

Varying conditions of total dissolved and suspended solids and pH.

High heat exchanger skin temperatures which may reach 145 deg.C. This may in some cases present some special demands on the additive. When applicable, the treatment should help to minimise the effects of nucleate boiling, reducing boiler noises and the resultant possible damage to the system.

The protective layer on the metals should be able to resist high water velocities and be equally protective under stagnant water conditions.

Whilst the treatment should exhibit good wetting properties, it should not encourage leaks from previously water-tight joints. The treatment should not support or promote foam conditions in areas of high turbulence.

It should possess some dispersing properties and it must be able to protect metal surfaces under the sediment of existing oxide sludge and other debris. Likewise, it must penetrate engineering crevices.

The water treatment should be effective over the total temperature range and must continue to provide protection even under conditions of prolonged plant shutdown.

The water treatment should also demonstrate the following characteristics and functions:-

It should be neither merely cathodic nor anodic in action but achieve a good working balance of both. The resultant protective layer must be fairly uniform, suitably overlapping, densely packed and non porous.

Absorption and chemisorption of the inhibitor should not significantly deplete the treatment's reserve in the circulating water, instead it should come to an abrupt end when the last layer of film is identical to the inhibitor.

Protective films must not crack or craze at points of high skin temperature and these films must not interfere with the engineering dimensions or thermal properties of heat exchangers.

The inhibition control should be effective under conditions of overdilution or overconcentration and it should protect all the metals likely to be present in the system.

The water treatment should be cost effective and ideally possess sufficient reserves so that precise additions are not necessary and, where water losses are negligible, only require occasional topping up. Such reserves would also allow for the treatment to minimise the effects of erosion corrosion where caused by the build up of suspended solids in evaporative systems, as well as compensating for any adsorption of inhibitor on the debris.

The treatment should not be aggressive to other engineering materials including solders, jointing compound, rubber and plastic components.

The water treatment should not give rise to the establishment or growth of organic impurities (fungi, bacteria, yeasts etc.), but should possess some micro-biological control as it may be added to existing polluted installations.

The treatment should prevent the precipitation and deposition of hardness salts and not form any precipitates on its own or in conjunction with hardness salts or other impurities. This is most important as pumps, valves water and heat meters may all be damaged by small particulate matter entering the water lubricated shaft/

bearing assembly. In fact ideally the treatment should help to reduce friction in such water lubricated parts.

Where necessary the treatment should be compatible with freezing point depressants (or be available in the form of a non-toxic inhibited anti-freeze) for use in central heating systems, chilled water systems or solar heating installations. In addition it should have sufficient ability to counteract oxidation of the freezing point depressant, which occurs naturally with time and temperature, thus helping to extend the life of the antifreeze.

It should be non-volatile. The concentrate should be non-toxic and non-irritant and should not provide any other health hazard. It should be easily applied, and the resultant treated water should also remain non-toxic and be environmentally friendly.

Finally both chemical spot tests and performance tests should be available so that the efficiency of the water treatment can be measured in the field.

4 A CLASSIFICATION OF INHIBITORS

Based on many years observations and tests on available products, we have categorised inhibitors according to performance, under the following headings:-

A "Safe Inhibitor" should not only perform to the criteria already outlined but should provide an in-use solution within the range of pH 5-9 and must reduce or retard corrosion uniformly.

An "Unsafe Inhibitor" may reduce corrosion of one or more metals or alloys, whilst failing to protect, or even promoting attack on, others present in the system.

A "Dangerous Inhibitor", when used within 10 percent of the recommended dosage rate, can tolerate or cause localised corrosion whilst apparently protecting much or most of the rest of the metallic surfaces. This condition can lead to deeper pitting and earlier perforations than without additive.

A "Damaging Inhibitor", is one that when used at the recommended dosage is "Safe" but when overdiluted or overconcentrated by, say, 25 percent can tolerate or cause localised corrosion with the same results as a "Dangerous Inhibitor".

"Corrosion-Proofers" on the other hand prevent rather than 'reduce' or 'retard' corrosion, even in the presence of oxidising agents. Corrosion "Prevention" in this context should be understood to mean reaction products substantially less than 0.001 gm/cm2 becomimg detached from a sound metal surface over a period of 5 years in a sealed leak free heating system, or over a period of 14 months in a heating system with an open vented expansion device. A Corrosion-Proofer should be effective for an extended period and should, during this time, not require more than occasional topping up at intervals of, say, 10-20 years.

5 CHEMICALS USED IN WATER TREATMENTS

Corrosion inhibitors can be generally classified into anodic, cathodic or both and inhibition usually results from the inhibitor molecule being chemisorbed on the metal surface so forming a protective film either by itself or by combining with metallic ions. Alternatively the inhibitor may result in the metal forming passive oxide films, or the inhibitor may react with corrosion promoting species that are present in the system water. The type and level of inhibitor chosen will depend on system conditions. For example, it is advisable to passivate cooling systems with 5 times the normal daily dose for 5 days, and then follow standard threshold techniques.

Anodic inhibitors build a very thin and generally invisible protective film initially at the anode which may eventually spread all over the metal surface. Whilst the degree of protection offered by many anodic inhibitors is high, care must be taken to ensure that they do not fall below their respective critical concentrations as partial protection may occur leading to the promotion of pitting attack on the metal surface.

Cathodic inhibitors are generally less effective than the anodic type and form generally visible films along the cathode surface. Whilst accidental low concentrations below their critical concentrations are generally found not to increase localised corrosion attack, care must be taken that overconcentration does not occur as this can cause corrosive conditions in some cases.

Ambiodic corrosion inhibitors possess properties of both Anodic and Cathodic inhibitors, although the number of examples of this type of inhibitor are limited. More commonly this balance of protection is achieved by blending various types of inhibitor to achieve an overall result.

In general terms, if the mean corrosion rate is reduced to 25 mdd for all the metals present and there is no sign of pitting attack then an effective corrosion protection has been achieved although it is possible to achieve results orders of magnitude better than this, cost effectively. Satisfactory protection can be said to be achieved if corrosion has been reduced to 250 mdd or less, again with no localised attack. Corrosion levels above 250 mdd or exhibiting localised attack are generally found to be inadequate.

The following chemicals are commonly used for their water treatment properties:

ORTHOPHOSPHATES - These are used as either the mono, di or tri metal salts depending upon the requirement for pH control. They are anodic inhibitors but are rarely used alone as they form insoluble precipitates with water hardness salts. Phosphates should be used with care especially with aluminium heat exchangers as they can induce pitting attack under high heat flux conditions. Several European countries prohibit the discharge of phosphates into the sewage system.

POLYPHOSPHATES - These are complex linear condensed or polymerised phosphates with variable chain length. They are used as the metal salts of hexametaphosphate, tripolyphosphate and pyrophosphate. They all act as Cathodic inhibitors when in combination with calcium and protect most metals by an electrodeposition mechanism. Threshold levels of these chemicals will not only provide corrosion protection to the system but will also act as an effective dispersant for water hardness salts. Being environmentally acceptable at low levels polyphosphates can therefore be cost effectively used for the treatment of tuberculation in potable water supply lines and once through cooling systems where only low corrosion rates are experienced. It should however be remembered that polyphosphates will hydrolise under acid or alkaline conditions or under the influence of high heat flux and so revert to Orthophosphate and act as above.

SILICATES - These again are available as metal salts with various degrees of silicate polymerisation. Again at normal in-use levels they are environmentally acceptable. They are generally only used for once through systems and are synergised with other treatments such as polyphosphates. The efficiency of silicate anodic protection is less than that for polyphosphates with levels in the order of 100 ppm being required to afford inhibition. The silicates are stable only at relatively high pH and readily form insoluble salts with water hardness formers. These precipitates can be highly abrasive and can also form high thermal insulating layers on heat exchangers even when only forming thin films. They are, however, usefully employed in neutralising the effects of acidic soft cuprosolvent waters and can be used to minimise the dezincification of brass.

TANNINS AND LIGNINS - are present in a variety of forms and have been widely used for many years as either oxygen scavangers, crystal modifiers and sludge dispersants and cathodic film formers. Generally they are not used in isolation but are synergised by other additives. Special attention has to be exercised in the control of pH with many of this family, otherwise precipitation can result in rapid depletion of the actives present for use. Unlike many inorganic anions, lignins and tannins do not promote localised attack when present at below critical concentration levels. High levels of tannins and lignins have been observed to form thick films on heat exchanger surfaces under conditions of high heat flux and these films can become detached as large platelets which can cause pump and valve problems in heating systems.

BENZOATES - are another example of an organic inhibitor that does not promote localised corrosion when used below its critical concentration. Used usually as the sodium salt it may be termed loosely as an anodic inhibitor but is rarely used in isolation and is more commonly combined with nitrites both in heating and cooling systems. Benzoates are readily biodegradable and require regular additions and close control of biocides.

NITRITES - may be classified as anodic inhibitors and are used as metallic salts. Care must be taken that nitrite levels are maintained correctly as at lower concentrations or in the presence of high chloride or sulphate levels, selective attack on brasses and copper alloys may occur. Nitrites are normally synergised with alkaline buffering agents and other inhibitors and rarely used alone. Care must be taken not to combine Nitrites with organic amines, amides or other organic inhibitors that may dissociate under heat transfer or catalysed conditions to form nitrosamines or other suspected carcinogens. Nitrite solutions are readily degraded by various micro-organisms which restricts their ease of use in open cooling systems.

NITRATES - These salts are becoming increasingly used for the protection of solder and aluminium components, and are generally synergised with other inhibitors. Whilst not as problematical as Nitrites, care should be taken as to their use for the same reasons.

CHROMATES and **DICHROMATES** These are one of the most widely effective inhibitor combinations available and have been widely used in both heating and cooling systems. However due to their toxic effect on the environment their use is being discouraged and considerable work has been undertaken to find other additives with which to synergise so allowing lower levels to be effectively used. Whilst they are used as Anodic inhibitors, Chromates are susceptible to variations in aggressive anion concentration and care must be employed in maintaining their recommended dosage levels.

ZINC SALTS are widely used as Cathodic inhibitors especially in cooling systems. Zinc is rarely used alone but is synergised with other inhibitors so as to minimise the amount of zinc present in the system water. Zinc salts are considered to be environmentally toxic and their use is discouraged at more than low ppm levels. Close control of pH is necessary where zinc salts are used as they are liable to form precipitates in alkaline conditions. This may be improved by the addition of threshold agents.

MOLYBDATES are now being increasingly used as anodic inhibitors especially because of their low order of toxicity. Molybdates are being used as alternatives to zinc, chromium and nitrite as inhibitors. Due to the high cost of molybdates, they are not used alone, but synergised with other inhibitors. It should also be noted that molybdates are most efficient where dissolved oxygen levels are substantial and so they are ideally suited for use in cooling water systems. Molybdates however have only been used with limited success in open heating systems and severe limitations have been observed in molybdate inhibitor formulations used in pressurised heating systems where dissolved oxygen levels are rarely above 1 ppm.

AROMATIC AZOLES such as Benzotriazole, Tolyltriazole and Mercaptobezothiazole are widely used in inhibitor formulations as copper inhibitors.

PHOSPHONATES are now being used increasingly as cathodic corrosion inhibitors and as deposit control agents. They act in a similar manner to polyphosphates but are more resistant to hydrolysis and thermal breakdown under high heat transfer conditions. They usefully synergise with other inhibitors and can stabilise formulation concentrates. However it must be recognised that over-concentration of phosphonate formulations, especially in soft water areas, can create conditions of cupro-solvency and promote attack both on copper and brasses. Failure to control this defect can lead to considerable problems when attempting to protect mixed metal systems especially where aluminium components are present under conditions of high heat flux.

AMINOCARBOXYLIC ACID SEQUESTRANTS - These have been widely used as scale prevention additives but because they can only act in stoichiomentric proportions with hardness salt formers such as Calcium and Magnesium, they are costly to employ except in long term treatments. These sequestrants however will combine with other metallic ions depending on system pH and care must be taken in controlling their in-use conditions for the same reasons as for the phosphonates. Whilst sequestrants of this type are considered to be biodegradable, concern has been expressed in some countries on their ability to sequester heavy metals and allow these to be carried to sewage farms and eventually into the potable water supply.

LONG CHAIN POLYMERS - These tend to be synthetics of varying chain length and polarity, specifically designed to improve sediment and precipitate dispersion as well as other threshold effects. Care must be taken in the control of in-use concentrations of these molecules as in many cases above a critical concentration they can create considerable reprecipitation problems both in heating and cooling systems. The thermal instability of some of these polymers can lead to the eventual formation of insoluble organic platelets that can lead to blockages in circulators and valves. At high concentrations some of the shorter chain polymers can act as sequestrants.

SULPHITES and **HYDRAZINE** have been commonly used as oxygen scavangers in steam boiler plant. More recently they have been used with limited success in closed heating systems. As a result of toxicological studies, the use of Hydrazine is being discouraged and Sulphites alone have been proven to promote anaerobic bacterial attack in sealed pressurised systems.

LONG CHAIN AMINES and **AMIDES** have been successfully used in steam condensate lines but more recently they have been employed in conjunction with other inhibitors in closed and open heating systems. At lower temperatures this type of inhibitor has been found to be ineffective producing only partial protection for steel components and promoting pitting attack. The presence of corrosion sludge leads to the rapid depletion of the inhibitor's active ingredients. This type of inhibitor has been found to be aggressive to the jointing compounds, and rubber seals commonly found in heating systems. These films have also been observed to promote stress cracking in some types of plastic pipes.

SOLUBLE OILS - These can be classified as anodic inhibitors. Whilst their use may be cost effective in the short term, in central heating systems they can promote unacceptable attack on rubbers and other sealing components, as well as providing an ideal breeding ground for organic pollutants.

MICROBIOLOGICAL CONTROL AGENTS - These are formulated either into the water treatment concentrate or are used by separate regular addition to the system waters to help maintain control over the growth of bacteria, fungi, yeasts and in some instances algae. Care must be taken in choosing the type of agent used as well as ensuring that the rest of the components in the water treatment package and variations of system conditions do not reduce the agent's effectiveness. The formulator must also be continually aware of the possibility of the build up of immunity by organisms to the control regime being employed, as well as the likely effects on the environment of blowdown or system discharge.

When formulating water treatments one or more of the above may be combined to provide an overall protection for the system. It should be noted that, especially where threshold treatments are to be employed, certain cómponents may be selectively depleted through adsorption on to corrosion deposits and other suspended or precipitated particles. Such depletions if unnoticed can promote rapid localised attack. Similarly, where treatments are to come into contact with metals under high rates of heat transfer care should be taken to ensure that the formulation has been adequately tested for the effects of possible thermal degradation and catalyic oxidation. Many formulations themselves support microbiological growth and in such cases biocides must be used.

In the evaluation of candidate water treatments it is essential that full testing is carried out. It must not be assumed that because a mixture of ingredients has performed satisfactorily in the field in one application that it can be transferred to another. All too often presumptions of this type have led to considerable problems in both heating and cooling systems. Whilst short term electrochemical tests can be of value as a first line screening stage, long term tests both in the laboratory and in closely controlled field tests is strongly advised. Tests should contain both weight loss and electrochemical techniques as well as final microscopical inspection of the test specimens. Wherever possible testing of stressed and galvanically coupled metals should be included and the test coupons used should be of the same alloy and surface area ratios as found in the field.

Finally it is also essential that the formulation concentrate has a long shelf life. Liquid formulation concentrates should not recrystallise or undergo phase separation on freeze-thaw testing. The activity of the ingredients should not be depleted with time and the concentrate should not be susceptible to biodegradation. When considering treatment for an installation the formulator should be aware of the possibilities of feed water pretreatment to control dissolved gas levels, water hardness, dissolved solids and pH.

An ideal treatment programme therefore may often entail an understanding and balancing of various factors such as chemical, energy and maintenance costs as well as a capital plant expenditure.

By successfully considering all these points, the discerning formulator can provide a service best suited to his customer's requirements in the most cost effective way.

Mechanisms of Corrosion Inhibition in Neutral Environments

By S. Turgoose

CORROSION AND PROTECTION CENTRE, UMIST, PO BOX 88, MANCHESTER M60 1QD, UK

1 INTRODUCTION

A corrosion inhibitor may be defined as a substance that when added to a corrosive environment in small quantities reduces the corrosion rate of a metal. It is important when classifying a substance as inhibitor that both the metal and the environment are defined, and this also applies to discussion of inhibitor mechanisms. The discussion in this paper is restricted to oxygen containing environments since deaerated near-neutral environments, such as may be found in carbon dioxide containing systems, give different corrosion mechanisms and require the use of different inhibitors.

Classification of Inhibitors

There have been many classifications of inhibitors proposed: e.g. adsorption or film-forming; anodic, cathodic or mixed; safe or dangerous; oxidising or non-oxidising; organic or inorganic. These are not mutually exclusive and some give no information about the mechanism of inhibition. The adsorption or film-forming classification does indicate how the corrosion rate is reduced but is rather general. In neutral solutions the vast majority of inhibitors are film-forming but these films can be formed, and can affect the corrosion rate, in different ways. Perhaps more useful from a mechanistic view is the division of inhibitors according to whether they affect the anodic process, the cathodic process, or both (mixed inhibitors). It should be stressed that it is not necessarily correct to classify a substance as a particular type of inhibitor, since some can act in different ways depending on, for example, the inhibitor and aggressive ion concentrations. Chromate is an example of this as will be discussed later.

The classification of inhibitors as anodic or cathodic is often based upon their effect on the corrosion potential of the metal, a rise in potential after addition of the inhibitor indicating an anodic effect and a fall showing a cathodic effect. This is informative with inhibitors that affect only one of the electrode processes but with inhibitors that affect both electrode reactions the potential can vary in either direction. The potential change does not necessarily indicate whether the anodic or cathodic effect is primarily responsible for the reduction in corrosion rate. Thus, it is possible to have an inhibitor that reduces the corrosion rate due to its effect on the cathodic reaction but which causes the potential to rise due to its effect on the anodic.

Inhibitors can reduce corrosion rate in one of two ways; by reducing the rate of the rate-determining step in the corrosion process, or by causing the introduction of a new rate-determining step. Thus discussion of the mechanisms by which inhibitors work needs to be based upon an understanding of what controls corrosion rate in the absence of the inhibitors. This depends to some extent on the metal, and the following discussion is mostly concerned with iron and mild steel.

2 INHIBITORS FOR IRON AND MILD STEEL

In near-neutral solutions (pH 5-9) the predominant cathodic process in the corrosion of iron and mild steel is oxygen reduction, equation 1. In the absence of inhibitors

$$O_2 + 2H_2O + 4e = 4OH^- \qquad (1)$$

the rate of corrosion of these metals is controlled by the rate of oxygen mass transport to the metal surface, as indicated in the E-log i diagram, Figure 1.

Anodic Inhibitors

Anodic inhibitors are ones that cause the formation or maintenance of a passive film on the iron surface and this film essentially prevents the anodic dissolution of the metal. They are also termed passivating inhibitors and passivity is accompanied by a high corrosion potential (commonly in the range 0 to -250 mV sce). Anodic inhibitors are often further classified as oxidising or non-oxidising

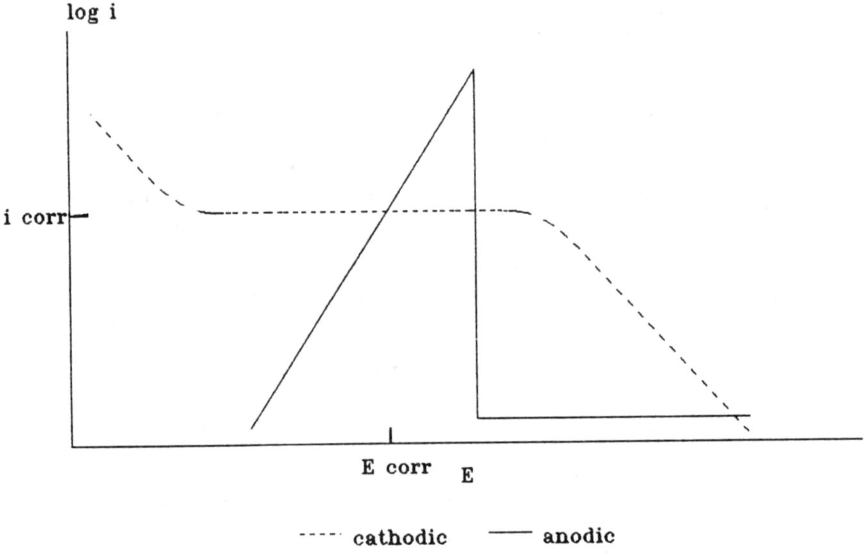

Figure 1 E-log i diagram for iron in neutral solution.

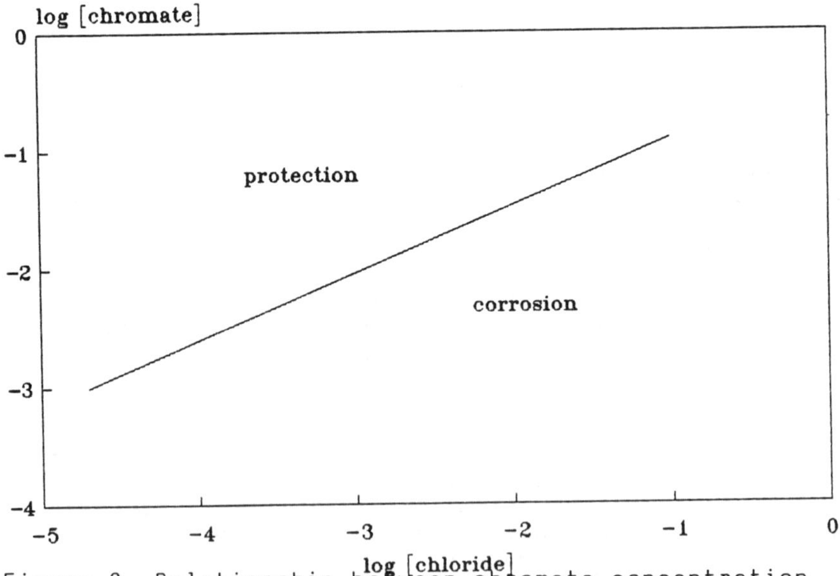

Figure 2 Relationship between chromate concentration required for passivity of iron and chloride concentration.

on the basis of whether they require oxygen in the solution in order for passivity to develop. There has been much work over the years concerned with the mechanism of action of anodic inhibitors and most of this work has been reviewed in standard texts.[1,2,3] The most common oxidising inhibitors are chromate and nitrite and typical non-oxidising inhibitors include phosphate, molybdate and many carboxylates, e.g. benzoate.

It is important in a mechanistic discussion that the various properties of anodic inhibitors should be explained by any proposed mechanism, especially the effects of environmental parameters such as inhibitor concentration, aggressive ion concentration, and solution pH. With anodic inhibitors there is a minimum concentration required for passivity, and below this concentration there is no reduction in corrosion rate. The minimum concentration depends on the concentration of aggressive ions, such as chloride, sulphate and nitrate. A linear relationship exists between the logarithm of the inhibitor concentration required for protection, C_{inh}, and the logarithm of the aggressive ion concentration, C_{agg}, as given in equation 2, where n is often close to the ratio of the charge on the inhibitor to the charge on the aggressive ion. This

$$\log C_{inh} = n \log C_{agg} + \text{constant} \qquad (2)$$

relationship has been demonstrated for chromate, nitrite and benzoate as inhibitors,[4,5,6] and that for chromate-chloride mixtures is shown in Figure 2. There is also a minimum pH below which protection is not achieved, and this may be dependent on the concentration of the inhibitor.[7,8]

Spontaneous passivation of an active surface requires that the limiting cathodic current density (i_{lim}) should be greater than the critical current density for passivation (i_{crit}), and this indicates two ways in which passivation may be caused, by reducing i_{crit} or by increasing i_{lim}. These are shown schematically in Figure 3, line 1 showing increased cathodic current and line 2 reduced i_{crit}, both leading to passivity. Some aromatic nitro-compounds which can be readily reduced at an iron surface, such as nitrobenzoate, do considerably increase i_{lim}, enabling passivation even with large values of i_{crit}.[3] Whilst these do not cause passivity solely by their effect on the anodic reaction the result of their action is the same as with conventional anodic inhibitors. However, most of these, even oxidising ones such as

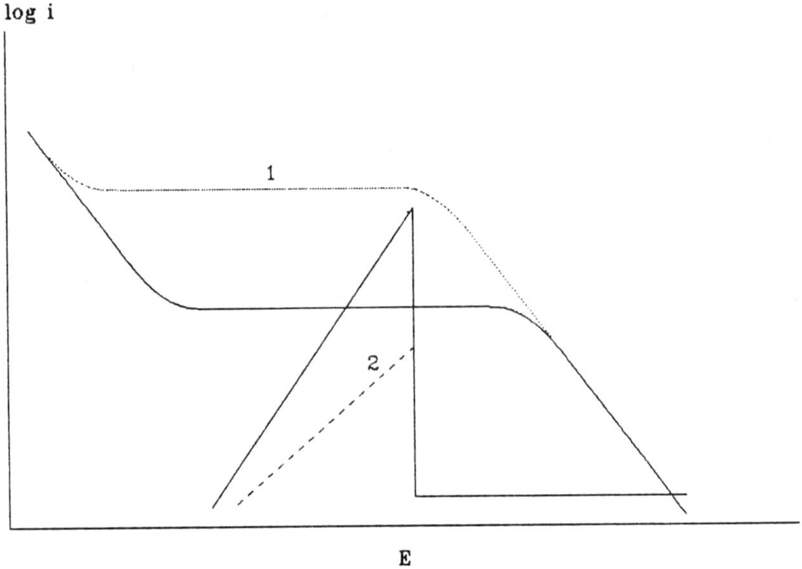

Figure 3 E-log i diagram showing passivation by increasing i_{lim}, (1), or decreasing i_{crit}, (2).

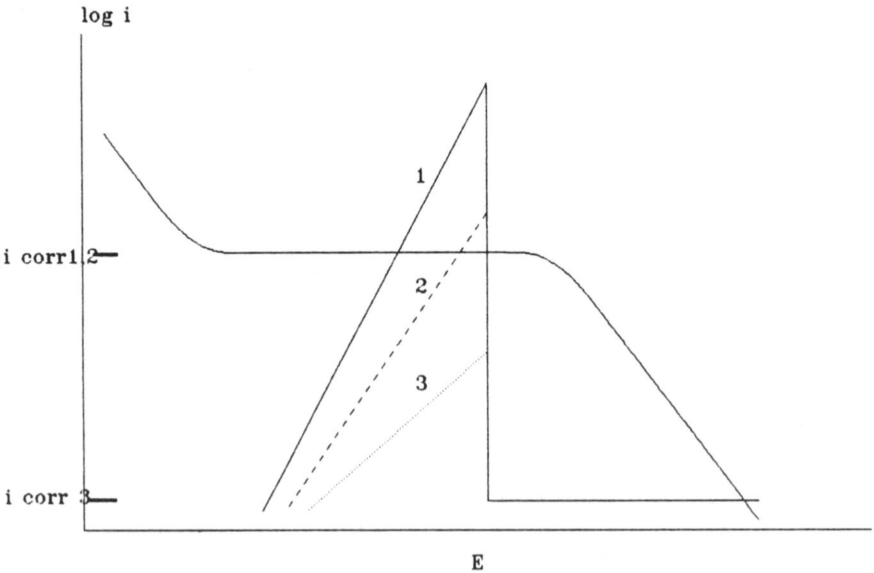

Figure 4 Effect of increasing anodic inhibitor concentration. Concentration increases from 1 to 3.

chromate and nitrite, rely predominantly on reduction of i_{crit}. The difference between oxidising and non-oxidising inhibitors is that the former can provide a cathodic reaction at passive potentials in the absence of oxygen whereas the latter can not. The reason for the abrupt change from no protection can be seen from Figure 4, the corrosion rate being determined by the cathodic reaction until the inhibitor concentration is sufficient to passivate the surface.

Studies of the mechanism of action of passivating inhibitors have concentrated on either the spontaneous passivation process, the requirements for which are described above, or on the ability of the inhibitor to maintain an existing oxide film on the surface. Consideration of this requires an understanding of the processes that occur during the breakdown of oxide films on iron in the absence of inhibitors.

The final stage of the breakdown is well established, and occurs by a process of reductive dissolution, where the reduction of the iron oxide, equation 3, is coupled to the normal anodic dissolution of the metal, equation 4.

$$Fe_2O_3 + 6H^+ + 2e = 2Fe^{2+} + 3H_2O \qquad (3)$$

$$Fe = Fe^{2+} + 2e \qquad (4)$$

The initial breakdown can be considered to occur by pitting. At local defects in the oxide film metal dissolution occurs, with the cathodic reduction of oxygen (and possibly other species) on the passive surface. As the pits grow the current required is greater and the potential falls. If the stage is reached where the potential falls sufficiently for oxide reduction to occur then the metal will activate. Thus pitting is only apparent in the early stages of the potential decay and pits can be clearly seen on samples removed before the general activation of the surface. Various theories for the action of inhibitors invoke their action either in preventing the formation of the anodic sites, or in preventing metal dissolution at sites that come to exist. The criteria given above for the spontaneous passivation of an active surface clearly also apply to the repassivation of small active sites except that the process would be expected to be easier in the latter case since the general requirement is given in equation 5, where A_c

$$i_{lim} \cdot A_c > i_{crit} \cdot A_{an} \qquad (5)$$

and A_{an} are the cathode and anode areas respectively. Since, with only a few anodic defects, $A_c >> A_{an}$ passivation can occur with $i_{lim} < i_{crit}$. It is commonly true that passivity is easier to maintain than to produce.

The currently available theory which seems to account best for both the maintenance of passivity and the active-passive transition is the "pore-plugging" theory. Strictly, this name relates to the maintenance of passivity by the blocking of anodic areas by the deposition of insoluble iron (usually ferric) compounds, but deposition of the same compounds can also be considered as responsible for the reduction in anodic current density in the active region.

There is considerable evidence to favour the pore-plugging theory. Apart form the fact that the known inhibitors give insoluble salts when mixed with ferrous ions (oxygen is also required for this with some of the non-oxidising inhibitors), small inclusions of ferric phosphate have been identified in oxide films on iron inhibited in phosphate solutions[9], the uptake of carboxylates such as benzoate and azelate has been shown to be localised[8,10] and the amounts of inhibitor taken up by surfaces are generally equivalent to less than monolayer coverage. The existence of the minimum pH for inhibition by benzoate, and its variation with the benzoate concentration, has been related to the relative stabilities of a basic ferric benzoate and iron oxide[8,11].

The localised uptake occurs even in distilled water indicating that the breakdown of the film occurs even in the absence of aggressive ions, which is not accounted for in theories based upon the ability of the inhibitors to adsorb and displace aggressive ions from the oxide surface and so prevent breakdown. The effect of aggressive ions can be explained if the solution chemistry within the small anodic pits is considered. If it is assumed that continued dissolution will occur above a certain aggressive ion / inhibitor ratio then allowance for the migration of these ions under a potential difference will give the observed logarithmic relationship between the bulk concentrations.

Cathodic Inhibitors

In this case the mechanism of inhibition is more clear although the reasons for the presence or absence of inhibitive effects in different systems is not always apparent. In aerated near-neutral solutions the rate of the corrosion is controlled by the rate of oxygen reduction. Cathodic inhibitors reduce the rate of this by forming thick (compared to passive films) films on the metal surface which reduce oxygen transport to the surface, reducing the corrosion rate as shown in Figure 5. The film must, of course, be electronically non-conducting otherwise the cathodic reaction would occur on the surface of the film.

The possible processes giving rise to insoluble products on the metal surface are:

(a) reaction of solution species with hydroxide to precipitate a solid,
(b) reaction of solution species with ferrous ions,
(c) reduction of solution species at the active metal surface followed by precipitation, and
(d) decomposition of solution species followed by precipitation.

Often more than one of these processes will occur, particularly with combinations of cathodic inhibitors and it can be difficult to separate those processes responsible for the reduction in corrosion rate from those that are irrelevant as regards inhibition but occur nevertheless.

The first is the mechanism attributed to classical cathodic inhibitors such as Zn^{2+} which precipitates zinc hydroxide, or calcium and bicarbonate in natural waters which precipitate calcium carbonate. The possibility of these reactions can be calculated knowing the solution pH, buffering capacity and inhibitor concentrations, if the solubility product of the possible precipitates is known. Obviously the surface pH must rise sufficiently to precipitate the solids, and this pH rise occurs due to the hydroxide produced by the cathodic reaction, equation 1. The surface pH on a clean surface with oxygen reduction as the cathodic reaction can be calculated as a function of bulk pH and will rise to about 10 in unbuffered solutions, ignoring hydroxide precipitation reactions, as long as the bulk pH is above 4. In practice the presence of ferrous ions may limit this pH to about 9, but the precipitation of compounds such as zinc hydroxide can easily occur, even if the bulk solution is noticably acidic. Even when a film is formed on the metal it might be expected that the surface pH would be almost independent of barrier properties of the film, i.e of the corrosion rate, since as O_2 diffusion

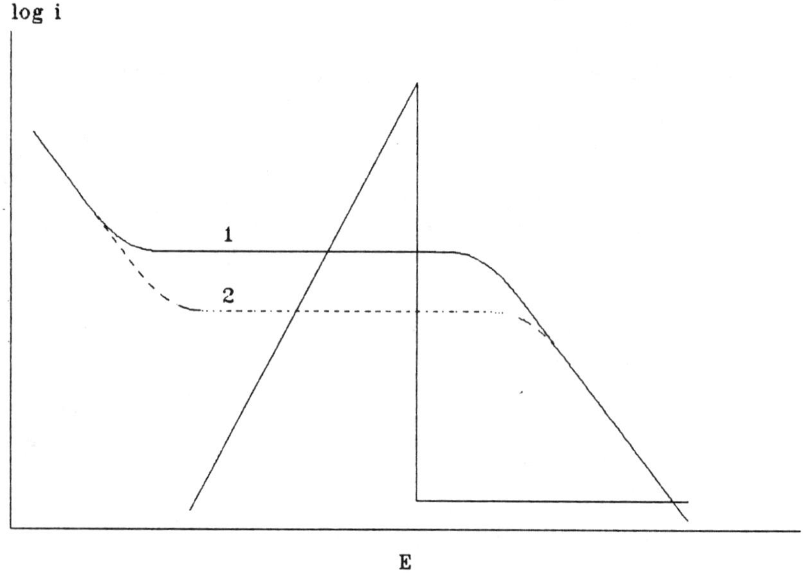

Figure 5 E-log i diagram showing effect of cathodic inhibitor. Without inhibitor (1) and with inhibitor (2).

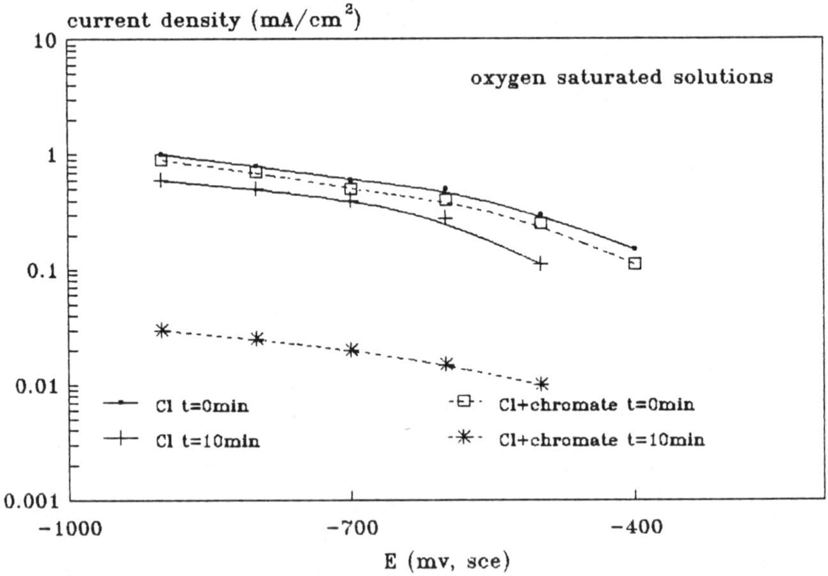

Figure 6 Potentiostatic polarisation of mild steel in 0.01 M NaCl with and without 0.00012 M chromate. Curves constructed from potentiostatic current-time measurements (from ref 13).

to the surface is hindered then OH^- diffusion away is also hindered. However, we must allow for the possibility that the films may not have equal permeability to neutral molecules and to ions.

Reaction with ferrous ions occurs in many systems. The simplest case is the reduction in corrosion rate due to rust formation, but with many inhibitors iron can be found in the surface film. It is often uncertain to what extent this contributes to inhibition, as opposed to being an inevitable consequence of the corrosion of the metal.

Reduction of the inhibitor is perhaps not too common, but does account for much of the inhibition by one important inhibitor, chromate, where reduction of chromate, equation (6) results in the formation of chromium (III) hydroxide or oxyhydroxide on the surface.

$$CrO_4^{2-} + 5H^+ + 3e = Cr(OH)_3 + H_2O \qquad (6)$$

That chromate can affect the cathodic reaction has been known for some time[3,12] and recent work has quantified the effects[13], Figure 6, and it has been demonstrated using XPS that chromium is present on the surface as chromium (III).[14] This is of importance since the conditions in most practical applications of chromate are such that a passive surface would not be expected. For example, a water containing 35 ppm (10^{-3} M) chloride would require about 1200 ppm (10^{-2} M) chromate to give passive steel, see Figure 2. Levels well below this are found to give effective inhibition and this must be attributed to the cathodic effect of chromate. The current decay curves in Figure 7, recorded on a rotating disc, show that even very low levels of chromate have a significant effect on the cathodic current under conditions of cathodic polarisation, and that the effect is rapid. Calculation of mass transfer rates to the surface of the disc suggest that the current has fallen to half its initial value in the time taken for the transport of sufficient chromate to give about a monolayer of Cr(III) hydroxide. At a freely corroding surface there is also the possibility of reaction of chromate with ferrous ion, resulting in a mixed iron(III)-chromium(III) oxide, but the presence of iron in the surface film is clearly not necessary for reduction of the cathodic current.

The decomposition of polyphosphates, to ortho-phosphate, with the subsequent precipitation of calcium

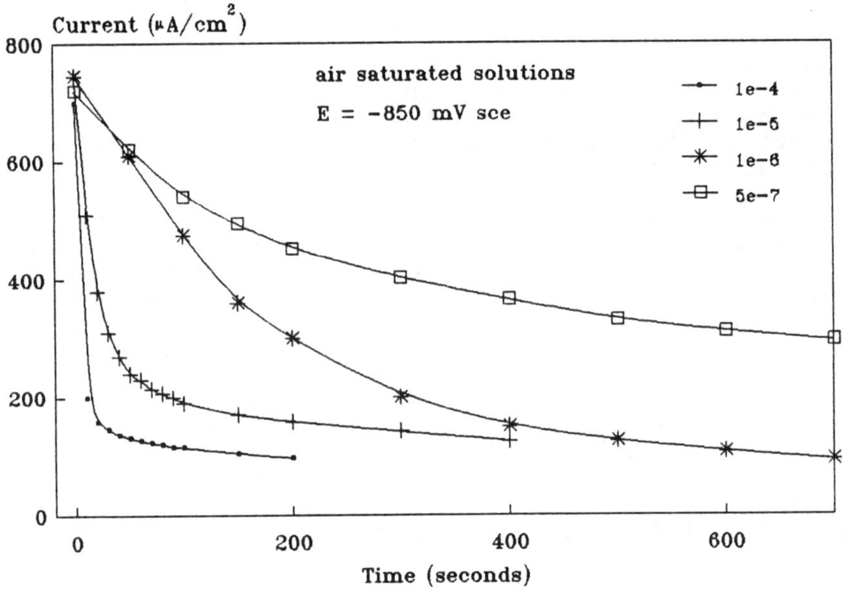

Figure 7 Current-time transients for mild steel in 0.1 M NaCl with the addition of chromate. Measured on a RDE at 1050 rpm. Numbers on curves are molar chromate concentrations.

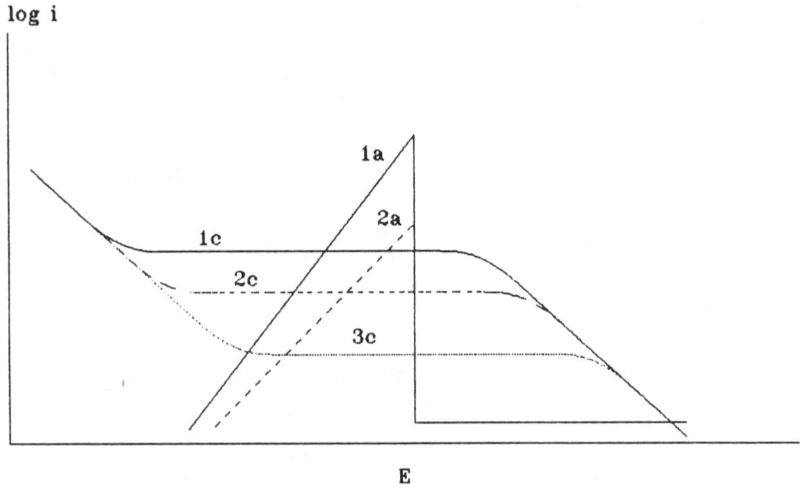

Figure 8 effect of "mixed" inhibitor with no passivation. No inhibitor (1); independent effects of inhibitors (2); increased cathodic inhibition if the anodic inhibitor also improves cathodic inhibition (2a and 3c).

phosphate has been suggested as the origin of inhibitive effects, although majority opinion seems to favour polyphosphate rather than orthophosphate films.

In many of these cases the nature of possible solids can be calculated from available thermodynamic data but in others the data is not available, especially for solubilty products of possible solids and stability constants of solution complexes. However, the simple formation of a solid at the metal surface is not sufficient to give significant reductions in corrosion rate. The surface films must be adherent to the surface and have low porosity. These properties may relate to factors such as the crystalline form of the precipitate, and the influence of various environmental parameters on this is not readily calculable.

Mixed Inhibitors

Many formulations, such as zinc with phosphate, chromate, molybdate etc. have been described as mixed inhibitors since under certain conditions the anions can act as anodic inhibitors and the zinc cations as cathodic inhibitors. However, this is not a particularly useful description since even with a mixture that affects the kinetics of both electrode processes the corrosion rate is still controlled by the rate of only one of the processes. The effect of separate additions of anodic and cathodic inhibitors on the corrosion potential and corrosion rate have been shown in Figures 4 and 5. The effect of joint addition with independent effects of the two inhibitors, for conditions where the anodic inhibitor does not passivate the surface, is shown in Figure 8, lines 2a and 2c. In this case the corrosion rate is still controlled by the cathodic rate, although the corrosion potential may rise, stay the same or fall depending on the extent of the effect on the anodic kinetics. This is different from acid solutions where the rate is determined by the activation polarisation of both electrode processes and where mixed inhibition is clearly beneficial.

However, it is clear that there are many mixtures that give corrosion rates lower than could be obtained with either component, and this must be explained. In general, mixtures used in applications such as cooling waters, have the "anodic" inhibitor present at levels insufficient for passivation, and the mixtures are in fact cathodic inhibitors. They are more effective cathodic inhibitors than zinc alone because the precipitated film is a basic salt rather than zinc hydroxide, giving rise to increased cathodic polarisation, as indicated by line 3c in Figure 8. This joint effect on the cathodic reaction has been shown for zinc-molybdate[15] and zinc-orthophosphate[16] mixtures. In the case of zinc and phosphate the possibility of surface film formation is predictable from the known solution chemistry of the system. This is discussed in more detail for this system in a later paper in this volume[19].

With zinc and chromate the inhibitive effect has also been attributed to the formation of a basic zinc chromate[17], but here the situation is slightly more complicated. Other work[14], using XPS, showed that all the chromium present was as chromium (III) although the nature of the technique involved the use of thin surface films and thus short immersion times. More recently, SEM investigations have shown that two different films are formed during immersion[18]. A thin apparently amorphous film forms across the surface initially, and analysis indicates a zinc : chromium ratio of less than 1 : 1. Later a crystalline layer grows over this and these crystals have a zinc : chromium ratio of about 5, as would be expected for the known basic zinc chromate, zinc tetraoxychromate, $Zn_5(OH)_8CrO_4$. Studies of the current time response at constant cathodic potentials would seem to indicate that the initial film is primarily responsible for the cathodic inhibition. However, there is no doubt that the reduction in corrosion rate, with low chromate levels and/or in the presence of aggressive ions, is due to a reduction in the cathodic reaction rate.

Not all examples of such synergistic effects are easy to explain. With chromate as a cathodic inhibitor the replacement of part of the chromate present by orthophosphate can improve the barrier properties of the cathodic film, and yet even very sensitive analytical techniques such as Auger spectroscopy can only detect minute traces of phosphorus in the surface film[14]. A possible explanation for this is that the phosphate modifies the crystal form of the chromium (III) oxy-hydroxide, so making it less porous. That additives can modify film morphology is well known in the area of conversion coatings, and since these cathodic inhibitor films form in exactly the same way, although much more slowly, it would not be surprising if such film modification occurred in many systems.

3 INHIBITORS FOR OTHER METALS

The inhibitors discussed above that function by affecting the mass transport of oxygen are also effective on other metals which have oxygen reduction as the cathodic reaction. These cathodic inhibitors are, however, much less effective in reducing hydrogen evolution from water which is a major cathodic reaction on some active metals such as aluminium and zinc. Possible reasons for this lack of effectiveness are that the film is not a barrier to water or that the hydrogen evolution destroys the adherent film.

The effectiveness of anodic inhibitors does, however, depend on the metal. In most cases effective anodic inhibitors are species which can form insoluble salts with ions of the metal being protected. For example, with zinc in the absence of inhibitors the corrosion rate is dependent on pH, showing a minimum over the pH range where zinc oxide is sparingly soluble. Inhibitors such as phosphate have insoluble zinc salts, which form surface films, and extend the pH range over which the corrosion rate of zinc is low. Similar arguments apply to lead, where inhibition by carbonate and bicarbonate can be attributed to the formation of surface films of lead carbonate, and to copper where effective inhibitors such as benzotriazole form protective films of copper salts on the surface. In view of the dependence of inhibition on the chemistry of the metal ion - inhibitor system it is not surprising that few anodic inhibitors can safely be used alone in multi-metal systems. The exception to this would appear to be chromate which has reasonable effectiveness on most metals. However, part of this could be due to the cathodic action of chromate and in this case a lack of specificity would be expected.

One interesting feature of the inhibition of active metals is that an inhibitor that only directly affects the anodic reaction may still reduce the corrosion rate even if the final inhibited rate is under cathodic control. This can occur if the effect of the inhibitor is to raise E_{corr} such that the rate of the hydrogen evolution reaction decreases, Figure 9.

There are some, at first sight, unusual occurrences with the use of anodic inhibitors for galvanic couples. For example, benzoate is only a moderate inhibitor for zinc alone and is very good for mild steel. It might therefore be expected that steel exposed at areas of damage on a galvanised surface would not corrode. In

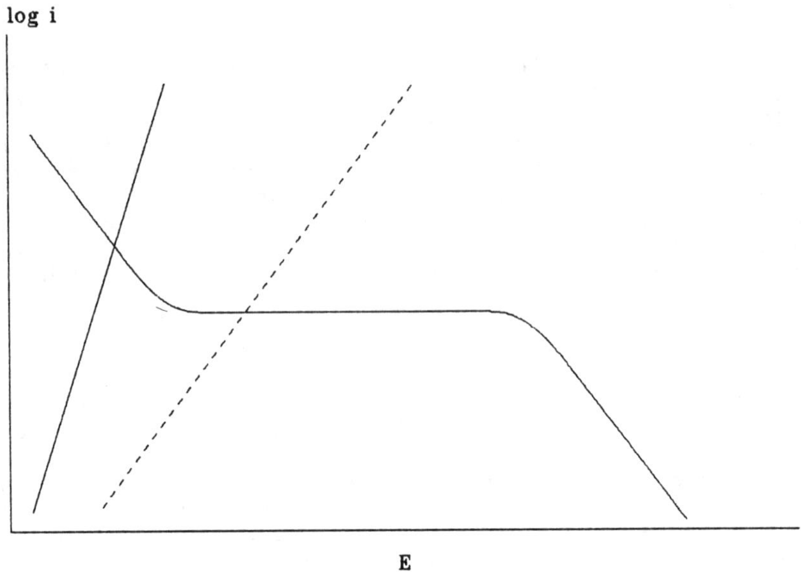

Figure 9 Effect of anodic inhibitor on the corrosion of an active metal.

fact the steel rapidly corrodes, after an induction period, protecting the zinc. However, this behaviour is predictable since coupling the steel to the zinc prevents the benzoate from passivating the steel. When a partially protective film develops on the zinc the polarity of the couple reverses. Another unusual effect, for which there does not appear to be an explanation at present, relates to the lack of effectiveness of benzoate, again, as an inhibitor for cast iron. Given that benzoate is an anodic inhibitor for steel, and the anodic reaction on cast iron is the same, this seems surprising.

4 CONCLUSIONS

As a general conclusion it would appear that the majority of inhibitive effects and environmental effects upon inhibitor action can be explained with current knowledge, although improved data concerning the solution chemistry of metal - inhibitor systems is required in many cases to enable quantification of the general conclusions that can be drawn. It should be stressed that, in attempting to predict inhibitor behaviour, it is the environment at the metal surface that is important and this can significantly differ from the bulk environment.

However, there are aspects of the inhibition process that are not currently amenable to quantification, especially concerned with cathodic inhibitors. Even if all the required thermodynamic data were available, so that the possibility of forming a surface film could be accurately predicted, there still remains the question of what determines whether a film that forms on a metal surface will be protective or not, and is it possible then to modify and improve non-protective films?

5 REFERENCES

1. L.L. Shreir (ed), "Corrosion", Newnes, London, 1976.
2. U.R. Evans, "Corrosion and Oxidation of Metals", Edward Arnold, London, 1960; 1st suppl. vol., 1968; 2nd suppl. vol., 1976.
3. I.L. Rozenfeld, "Corrosion Inhibitors", McGraw-Hill, New York, 1981.
4. D.M. Brasher and A.D. Mercer, Br. Corros. J., 1968, 3, 120
5. A.D. Mercer and I.R. Jenkins, Br. Corros. J., 1968, 3, 130

6. A.D. Mercer, I.R. Jenkins and J.E. Rhodes-Brown, Br. Corros. J., 1968, 3, 136
7. D.E. Davies and Q. Slaiman, Corros. Sci., 1971, 11, 671
8. C.L. Page and J.E.O. Mayne, Corros. Sci., 1972, 12, 679
9. J.E.O. Mayne and J.W. Menter, J. Chem. Soc., 1954, 103
10. J.E.O. Mayne and C.L. Page, Br. Corros. J., 1972, 7, 115
11. S. Turgoose, 6th European Symposium on Corrosion Inhibitors, Ferrara, 1985, 1045
12. G.B. Hatch, "Corrosion Inhibitors (ed. C.C. Nathan)", NACE, 1973, 126
13. E.A. Rossi, M.Sc. Dissertation, UMIST, 1985
14. M.A. Hazzaa, Ph. D. Thesis, UMIST, 1987
15. Y.J. Qian and S. Turgoose, Br. Corros. J., 1987, 22, 268
16. S. Seyed Razi, unpublished results
17. J. Golden and J.E.O. Mayne, Br. Corros. J., 1978, 13, 45
18. T.K. Dewi, M.Sc. Dissertation, UMIST, 1987
19. S. Seyed Razi & S. Turgoose this volume

The Current Trends in Cooling Water Corrosion Inhibitors

By G. E. Ibbotson and A. Marshall

GRACE DEARBORN LTD., WIDNES, CHESHIRE WA8 8UD, UK

INTRODUCTION

Control of ferrous metal corrosion in cooling water has been achieved with a variety of inorganic salts[1]. These include chromates, zinc, phosphate and nitrite, used either individually or blended to produce cathanodic inhibitor products.

Such materials technically have provided a degree of corrosion protection in cooling systems which allows trouble free operation with optimum plant life. Recently, however, the desirability of discharging such materials into the environment has become increasingly unacceptable.

Indeed the more advanced formulations have incorporated organic materials to reduce, and in some cases eliminate, the inorganic content. Similarly the introduction of the extended phosphate technology to replace the more toxic zinc chromate approach has been seen without sacrificing performance levels. Zinc based treatments incorporating polymers and organophosphorus materials have allowed a lower metal content to be used and applied to a wider range of water qualities.

Both advanced zinc[2] and extended phosphate[3] technologies require incorporation of organics for the stabilisation of the inorganic phase, to promote inhibition film formation and to prevent deposition.

The application of all organic treatments has been restricted to high alkalinity/calcium hardness waters in the primary role of calcium carbonate scale inhibitors.

The use of chromate, zinc or phosphate corrosion inhibitor in these waters is generally not necessary.

This paper describes the development of a total organic treatment which can be applied in a wide range of water qualities.

The total organic corrosion inhibitor treatment utilises a novel polymer enhancer system which makes possible corrosion inhibition in water qualities where previous organic products were ineffective. Investigation of the inhibitor mechanism using electrochemical techniques has been carried out in the laboratory. Evaluation of the treatment performance to prevent deposition and corrosion on clean and pre-corroded surfaces has also been examined.

EXPERIMENTAL

Corrosion and Scaling Rate Determination

The rates of scale deposition and corrosion have been evaluated in a laboratory test rig which has previously been described[4] in detail. From a reservoir the cooling water is pumped through a heat exchanger, followed by a coupon chamber, through a cooling tower and returned to the reservoir. Incorporated into the rig are automatic make-up and blowdown facilities to control the cycles of concentration.

Potentiostatic Measurements. A rotating disc electrode (Oxford Electrodes) coupled to the potentiostat (Princeton Model 363) was used. The experimental procedure for this technique has previously been described in detail[2].

The polarisation curves for mild steel were determined in water containing 0.1M $NaClO_4$, and 100 mg l^{-1} methyl alkalinity (as $CaCO_3$) at pH 8.5. All calcium hardness levels ranged from 0 to 400 mg l^{-1} (expressed as $CaCO_3$).

Treatment Evaluation. The potentiostatic technique was used to evaluate the properties of hydroxyphosphonic acid (HPA) alone and in combination with a low molecular weight cationic polymer (CAT). It was demonstrated[6] that a combination of HPA and CAT greatly enhanced the performance of HPA along as a corrosion inhibitor particularly in low calcium containing waters. Different ratios of HPA to CAT were investigated for effects on corrosion rates.

Two traditional treatments, zinc chromate and extended phosphate have been compared with TOCI for scaling and corrosion rates under the same condition. The results are shown in Table 1.

In field applications the ability of a treatment system to perform in pre-corroded systems is important. Pre-corroded test specimens, heat exchanger and monitoring probes were achieved by operating the test rig for 2 days without treatment prior to introduction of TOCI.

Figure 1 shows the instantaneous corrosion measurements for this test and Table 2 shows the overall corrosion and scaling rates with and without pre-corrosion.

Table 1 Comparison with Traditional Technology

Product Type	Product Dose mg l^{-1}	System Water Calcium Hardness mg l^{-1}	System Water pH	Mild Steel Corrosion Rate (mpy)		Deposit -ion mg dm^{-2}
				Coupon	Heat Exchanger	
NONE	NIL		7.5	48.0	42.1	3000
NONE	NIL		8.5	5.6	8.0	400
ZINC CHROMATE	35		7.5	0.6	0.5	200
EXTENDED PHOSPHATE TECHN- OLOGY	100	360	7.5	0.4	0.6	10
TOCI	100		8.5	0.8	0.7	10

Table 2 Effect of Pre-Corroded Surfaces on the Performances of TOCI

Additive	Test Duration (Days)	Mild Steel Corrosion Rate (mpy)		Deposition mg dm^{-2}
		Coupon	Heat Exchanger	
NONE	20	42.1	48.0	3000
NONE	2	24.0	26.6	185
**TOCI	20	1.2	0.4	10

** Using pre-corroded specimens

RESULTS AND DISCUSSION

In low calcium hardness waters (below 300 mg l^{-1}) the incorporation of HPA alone does not produce sufficient corrosion inhibition of ferrous metals, as very high corrosion rates were obtained (Table 3).

Table 3 Effect of Calcium Hardness on Performance of HPA and the optimum HPA/CAT mixture

System Water	Mild Steel Corrosion Rate (mpy)	
	HPA	HPA/CAT
50	26	1.8
100	-	0.2
150	14	0.2
200	-	0.1
300	-	0.2
400	0.5	0.5

The corrosion protection afforded, however, is significantly improved when a combination of HPA/CAT is employed. Indeed over the range of calcium hardness studied the optimum HPA/CAT blend does provide an effective mild steel corrosion inhibitor system.

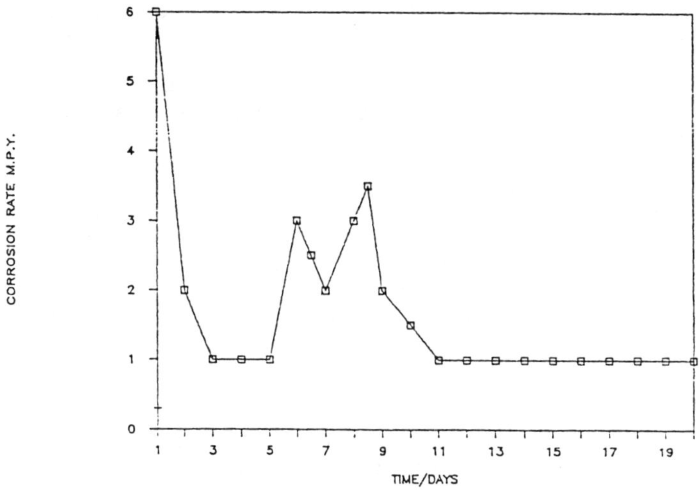

Figure 1 - Instantaneous Corrosion Rates for Rig Test Carried Out with Inhibitor Dose Upset.

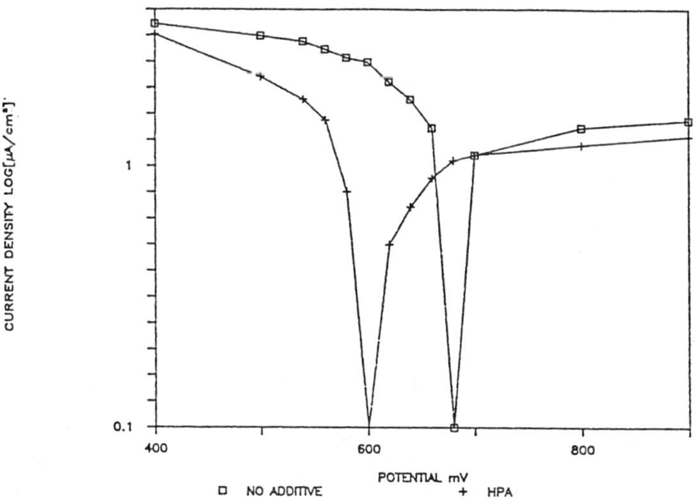

Figure 2 - Polarization Curves for Mild Steel in 0.1M $NaClO_4$ at pH 8.5 Both With and Without HPA.

Previously, HPA alone was studied in highly alkaline and high calcium hardness waters as a ferrous metal corrosion inhibitor. However, in such waters corrosion is not generally the primary problem, whereas in lower calcium hardness waters it is the major concern.

The discovery that the HPA/CAT system greatly enhances the range of water qualities over which effective corrosion inhibition is observed, is seen as a major breakthrough. This technology provides the basis of a low toxicity total organic corrosion inhibitor suitable for replacing the older and more toxic inorganic systems.

Mechanistic Studies

Having made the discover it was of interest to investigate the mechanism of corrosion inhibition.

The effectiveness of HPA as a corrosion inhibitor for mild steel can be gauged from the polarisation behaviour of mild steel in a medium such as sodium perchlorate.

Figure 2 shows log current against potential curves for mild steel in 0.1M $NaClO_4$ both with and without the addition of 30 mg l^{-1} HPA. The relative corrosion currents indicate a reduction in corrosion rate of about 50% in the presence of HPA. The cathodic polarisation curves indicate only a slight decrease in the limited diffusion current for dissolved oxygen. The major effect of HPA is to move the corrosion potential to less negative potentials and to reduce greatly the anodic currents which indicates that the HPA is primarily an anodic inhibitor.

Figure 3 shows the polarisation curves generated under the same conditiions as Figure 2 but with the addition of 300 mg l^{-1} of calcium hardness.

In the presence of calcium salts,(Figure 4) the cathodic polarisation curves indicate a greater decrease in the limited diffusion for dissolved oxygen and HPA behaves as a cathanodic inhibitor.

The formation of a calcium HPA complex on the metal surface is probably responsible for the increased cathodic protection in calcium containing waters.

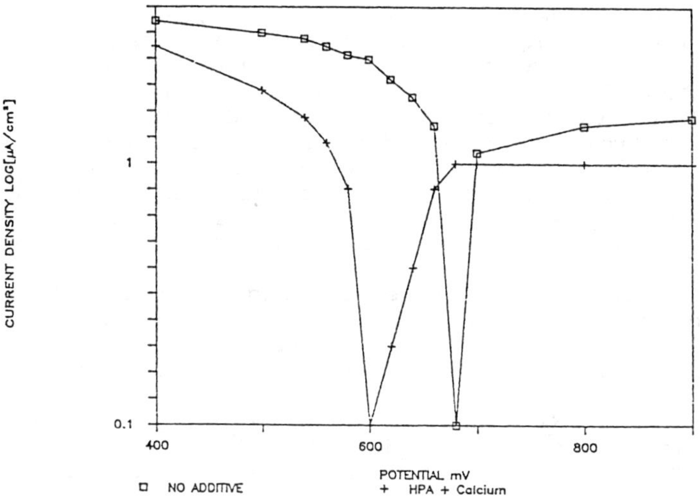

Figure 3 - Polarization Curves for Mild Steel in $0.1M$ $NaClO_4$ at pH 8.5, 300 mg/l Calcium Hardness Both With and Without HPA.

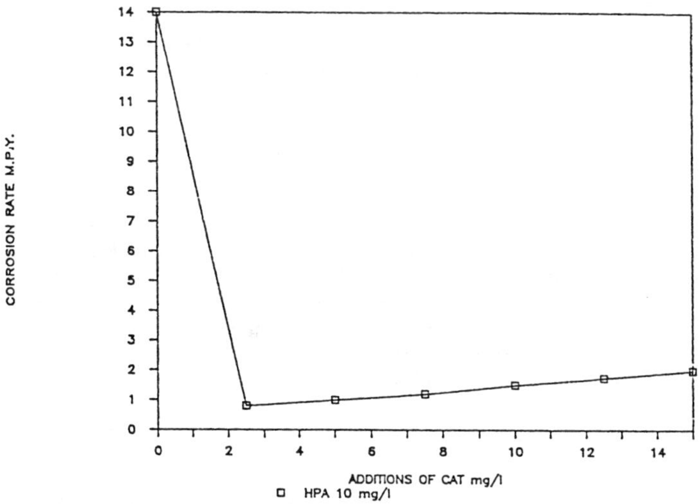

Figure 4 - Corrosion Rate of Mild Steel in the Presence of HPA/CAT Mixtures.

The CAT is not a corrosion inhibitor when evaluated alone, but as shown by Figure 4 can significantly improve the performance of HPA. The polarisation studies at a stationary electrode indicated that the major effect of CAT, when tested in conjunction with HPA, is to reduce the cathodic currents. The currents from the cathodic polarisation curves at -800 mV were plotted against calcium hardness concentration as shown by Figure 5. It can be seen that the CAT significantly enhances the cathodic inhibition provided by HPA in waters both with and without calcium hardness.

The mechanistic study suggests that the CAT is not an inhibitor, but plays a key role in transporting inhibitor, either HPA or calcim salts, to the metal surfaces which subsequently improves inhibitor film formation.

Test Rig Studies. The test rig was initially passivated with TOCI at 300 mg l^{-1} which after 2 days was reduced to 100 mg l^{-1} and then maintained at this level.

Figure 6 shows that the formulated TOCI gives excellent corrosion inhibition properties for the mild steel heat transfer surfaces over range of calcium hardness of 50 to 300 mg l^{-1}. At calcium hardness levels, lower than 50 mg l^{-1}, unacceptably high corrosion rates were seen. However, increasing the TOCI concentration to 150 mg l^{-1} as shown in Table 4, it is possible to achieve effective corrosion inhibition.

Table 4 Effect of Dose Rate on Performance of TOCI in low Calcium Hardness Waters

Product	System Water Calcium Hardness mg l^{-1}	Mild Steel Corrosion Rate (mpy) Heat Exchanger	Mild Steel Corrosion Rate (mpy) Coupon
100	25	21.9	1.9
150		1.4	0.8

In industrial situations the possibility of upset conditions is always present and it is of value to assess product performance under these conditions.

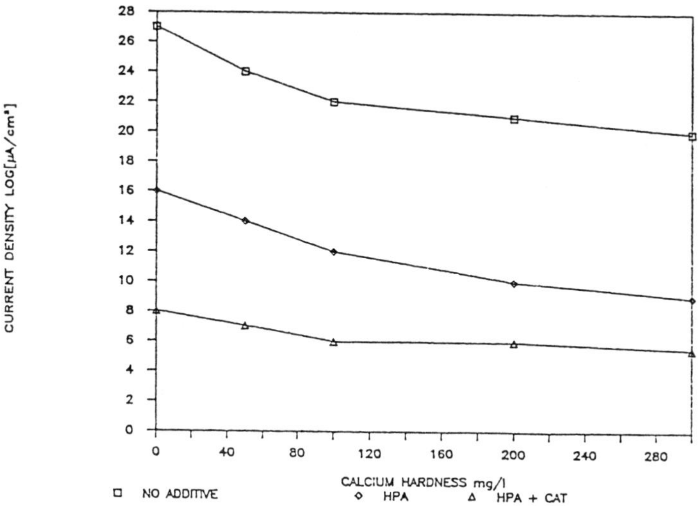

Figure 5 - Effect of Calcium Hardness on the Cathodic Current at -800 mV in the Presence of HPA and the Optimum HPA/CAT Mixtures.

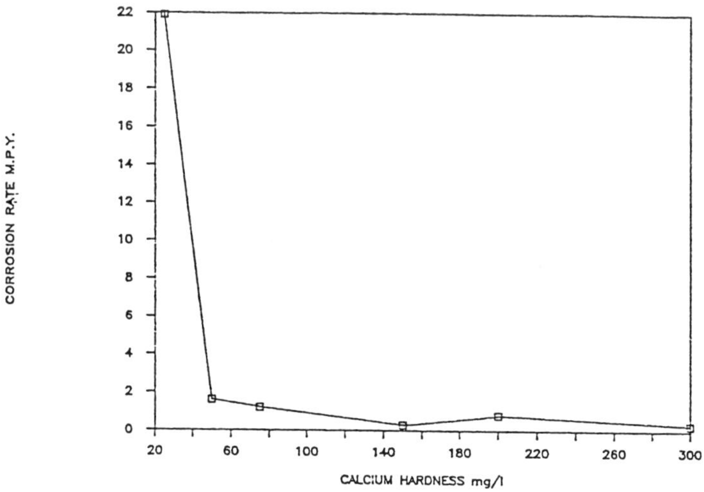

Figure 6 - Effect of Calcium Hardness on Mild Steel Corrosion Rates in the Presence of TOCI.

It is a feature of the more toxic materials such as zinc chromate that they can give rapid recovery after low pH or product dosage. Similarly for the recognised alternative, extended phosphate technology, satisfactory performances have been demonstrated in the area of upset conditions.

TOCI was therefore evaluated under upset conditions; these being reduced inhibitor dose level and depressed sytem pH, simulating acid spills.

Inhibitor Dose Level Upset. In order to study the flexibility of TOCI a rig trial was carried out with an induced low inhibitor level. A rig trial was run as normal, but after 6 days, and again after 8 days, a product excursion was carried out. In each case the level of product in the system was reduced to 50 mg l^{-1} instantaneously and allowed to build up to the normal 100 mg l^{-1} gradually with the concentrating effect of the rig.

Figure 1 shows the instantaneous corrosion rates measured in the test rig during the trial period. Corrosion rates increased three fold to 3 mpy with the initial upset and worsened slightly, to 3.5 mpy with a further reduced inhibitor level. This was followed by a rapid fall in corrosion rates when the reserves of inhibitor were returned to normal and within two days corrosion rates had returned to pre upset levels.

Table 5 Effect of Inhibitor Dose Upset on the Performance of TOCI

Product Dose mg l^{-1}	System Water Calcium Hardness mg l^{-1}	Inhibitor Dose Upset	Mild Steel Corrosion Rate (mpy)		Digestion mg dm^{-2}
			Coupon	Heat Exchange	
NONE		NO	48.0	42.1	3000
100	150	NO	1.2	0.4	10
100		YES*	1.5	1.2	10

* Dose rate reduced to 50 ppm on Day 6 and Day 8.

Comparison of the rig trial results both with and without the inhibitor dose level upset (Table 5) indicate that the overall corrosion rates on the heat exchanger remain virtually unchanged by the upset and the scaling rate remaining unaffected.

<u>System pH Upset</u>. The ability of TOCI to passivate mild steel surfaces after induced low pH was assessed. Although the likelihood of acid spills is greatly reduced, since the TOCI approach does not normally require pH control it was of interest to study the recovery rate from a low pH upset.

A test rig was operated under normal operating conditions for 8 days after which the system water pH was reduced to pH 4.0 by addition of sulphuric acid for a period of 4 hours. At the end of this period the pH was adjusted back to pH 7.5.

Corrosion rates measured by the instantaneous corrosion meter increase rapidly to 20 mpy in response to the low pH (Figure 7). However, even before the pH was restored the corrosion rate had reduced to approximately 10 mpy and within 24 hours of the upset corrosion rates had returned to 1 mpy.

During the pH upset the surfaces of the heat exchanger and coupons darkened gradually and some breakdown of the inhibitor film took place. On restoring the system pH, areas where corrosion had been initiated were quickly passivated.

<u>Table 6</u> Effect of pH Upset on Performance of TOCI

Product Dose mg l^{-1}	System Water Calcium Hardness mg l^{-1}	pH Upset	Mild Steel Corrosion Rate (mpy) Coupon	Heat Exchanger	Deposition mg dm^{-2}
HOME		NO	48.0	42.1	3000
100	150	NO	1.2	0.4	10
100		YES*	1.4	0.8	10

* pH reduced to 4.0 for 4 hours

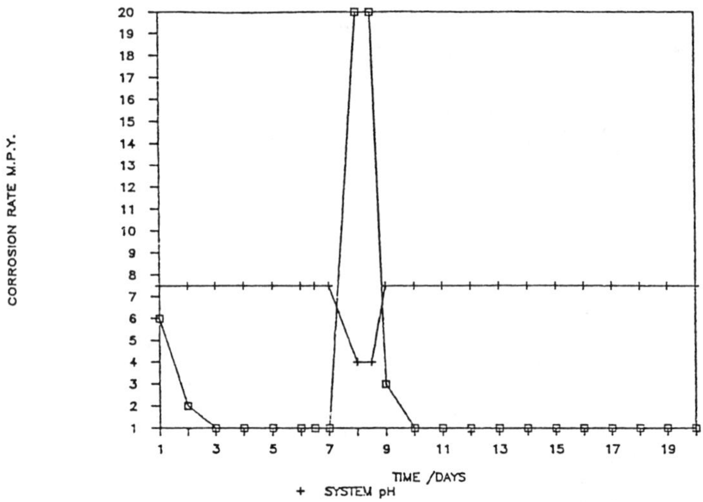

Figure 7 - System pH and Instantaneous Corrosion Rates for Rig Study Carried Out with Low pH Upset.

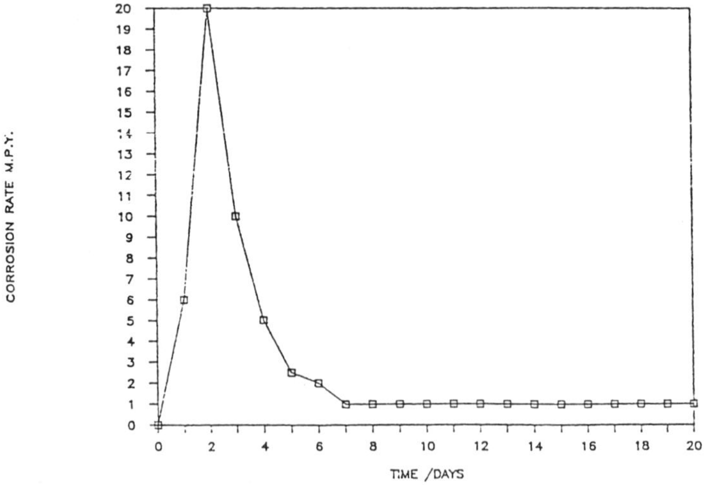

Figure 8 - Instantaneous Corrosion Rates for Rig Study Using Pre-corroded Specimens.

Comparison of the rig trial rsults both with and without the pH upset (Table 6) suggests that no serious damage had been inflicted on the heat exchanger surface by the pH upset, since the overall corrosion rates had only increased slightly and the scaling rate remained unaffected. Indeed no fouling with corrosion products was noted, this being consistent with the low level of overall metal loss.

Table 1 shows, the results of studies comparing TOCI with traditionally used cooling water treatments under upset conditions.

High corrosion rates of approximately 40 mpy were seen on both heat exchanger and test coupons in the blank test. Significant deposition of predominantly iron oxide occurred at the heat exchanger.

Introduction of zinc chromate product at 35 mg l^{-1} (yielding 15 mg l^{-1} of chromate and 2.5 mg l^{-1} of zinc) in to the circulating water did significantly reduce corrosion rates but some deposition still occurred. Extended phosphate technology achieved similar results but with better deposition control. However, TOCI achieved by corrosion inhibition and deposition control without the need for pH correction by acid dosing.

<u>Pre-corroded Ferrous Metal Surfaces.</u> Introduction of inhibitor programmes into systems with previously corroded ferrous metal surfaces is not uncommon. It is therefore of value to understand the effect on corrosion rates of such metal surfaces.

The study was achieved by allowing test pieces to pre-corrode in the test rig with water alone prior to the introduction of the inhibitor. This produced surfaces which were completely covered with corrosion products and measured corrosion rates after 2 days were about 25 mpy (Table 2). After 20 days and with significant levels of deposition occuring, corrosion rates increased further to over 40 mpy.

Introduction of TOCI reduced instantaneous corrosion rates of the pre-corroded probes from 20 mpy down to 1 mpy over a 4 day period as shown by Figure 8.

Examination of the pre-corroded specimens indicated that the treatment had arrested the corrosion occuring and reduced the rate from 25 mpy down to 5 mpy. Furthermore, the total amount of deposition on the heat exchanger was dramatically reduced indicating that TOCI may have some online cleaning properties.

CONCLUSION

The investigation demonstrates that hydroxy-phosphonic acid (HPA) alone is a corrosion inhibitor for mild steel. Incorporaton of a low molecular weight cationic polymer (CAT) improves the corrosion inhibition particularly in low calcium hardness waters.

The optimum mixture of HPA/CAT was used to form the basis of a formulated product which provides effective corrosion control in a wide range of waters and to be resistant to upsets.

The development of TOCI represents a major breakthrough in cooling water technology, since this is the first time that total organic treatment has been shown to give the same performance as the more toxic treatments such as zinc chromate.

ACKNOWLEDGEMENT

The Authors would like to thank Dr. C. Wederell for carrying out the electrochemical studies and W.R. Grace for giving permission for this paper to be presented.

REFERENCES

1. W.A. Hess, Corrosion, 1960, 16, 18.

2. A. Marshall, 'Anti-Corrosion', May 1984.

3. A. Marshall and B. Greaves, 'Corrosion 86', Houston, 1986, p.399.

4. A. Marshall and B. Greaves, 'The Development and Use of Corrosion Inhibitors', Oyez, London, 1983.

5. A. Marshall, Corrosion Prevention and Control, 1983, 30, 5.

6. Patent Application GB.

7. M. Yeoman and A. Harris, Corrosion 86, Houston, 1986, Paper 14.

Waterside Corrosion in Steam Raising Boiler Plant

By A. Atkinson

ATKINSON CHEMICALS LTD., 23 HIGH STREET, TWYFORD, READING,
BERKSHIRE RG10 9AB, UK

1 INTRODUCTION

There are several interesting facets of corrosion that can be observed in a steam raising boiler plant. The degree of troublesome corrosion encountered can often be related to the conscientious efforts of the responsible boiler house engineer, and a technical input and co-operation received from the water treatment service company's representative.

In practice there are several fundamental aspects of boiler plant operation that can greatly influence the corrosion activity experienced. A check of the boiler plant steam chart(s) will very often show a fluctuating pattern of steam output. The sharp peaks indicating sudden steam demands will often be acccompanied by a 'priming' situation, i.e. alkaline boiler water droplets entrained in the steam leaving the boiler. Further, the boiler steaming schedule can be 5 days a week for a daily 8 hour period to a continuous 24 hours a day 7 days a week from a boiler on full stand-by duty to being operated week on week off. Finally, a boiler that has been internally acid washed and not correctly passivated can give rise to the initiation of corrosion, which, is often aggrevated when the boiler is put into service.

The more common types of corrosion problems experienced in the low to medium pressure boiler plants will be discussed considering the installation in three sections.

- boiler feed systems
- boiler internals

- condensate system

2 BOILER FEED SYSTEM

The boiler feed system constitutes a deaerator (if fitted), condensate receivers, feed water tank, feed lines, pumps, valves, chemical dosing points etc.

The boiler feed tank is a very important integral part of the boiler plant where negligence (unawareness) of good design can greatly influence corrosion experiences in other parts of the plant.

2.1 Feed Tank Design.

Figure 2.1 is a schematic representation of a recommended feed tank design. The basic feature is to ensure good mixing of incoming cold water make up with any returned condensate. Further increase in feed water temperature can be achieved with steam injection or heating coils. A feedwater temperature of 80-90°C is the usual target for attainment. These simple design features help in de-aeration of the feedwater, Henry's Law, and avoids cold water stratification in the feed tank. The reduced dissolved oxygen and carbon dioxide concentrations reflects more benefits in other sections of the boiler plant, see later, and also reduces the chemical needs and financial outlays.

When adopting the improved feed tank design features and higher feed water temperatures it is important to be aware of:

2.1.1 Feed-pump Cavitation. Boiler feed pump cavitation can arise if there is insufficient pressure head on the pump to cater for the higher feed water temperatures.

2.1.2 Heating Coils. A frequent practice is to put a steam, or boiler blowdown water, heated coil in the feed tank to raise the feed water temperature to the desired value. When using this design feature care should be taken to avoid the fresh, cold and fully oxygenated water making an immediate contact on the hot coil, otherwise the coil, often contructed of mild steel, will last approx. 3 months. before it becomes perforated.

There are other supplementary problems in the feed tank that can lead to corrosion if left unheeded.

2.1.3 Chemical Dosing Point. Very often the feed tank is the location frequently adopted for the dosing of the water treatment chemicals. The examples selected are often encountered and could apply to other chemicals and dosing points.

The actual location selected for a particular chemical is not in question but more the engineering aspect. Very often the source of the problem is linked with the actual fitting of the chemical delivery pipe into the feed tank, feed line, boiler or where-ever. The mistake relates to the pipe connection being fitted flush to the tank side. Several products, used as part of the water treatment programme, can be alkaline or acidic. A typical acidic additive is sodium bisulphite solution employed to scavenge oxygen, see 3.1.1. This product is added to the feed tank or feed line and in the concentrated form as supplied will chemically dissolve the material of plant construction which is usually mild steel, or galvanised steel (see Figure 2.1.3).

Apart from the dissolution of the tank, the production of ferrous/ferric ions react with the bulk water alkalinity and precipitate the respective iron hydroxides. This 'sludge' then gives rise to underdeposit attack in the feed tank and secondly adds solids to the water which can deposit and contribute to further corrosion and heat transfer problems. Caustic on galvanised steel is a further example.

2.1.4 Tuberculation on Mild Steel. This can often be observed in boiler feed lines if the water treatment programme is not effective. It relates to the presence of dissolved oxygen in the feed water. On examination the 'TUBERCULES' are noted to have an exterior reddish brown colouration and if the nodule is broken off the underlying deposit is black. This corrosion mechanism can be self perpetuating due to the generation of acidic conditions (hydrogen ions) within the pit (see Figure 2.1.4).

The release of the hard nodules into the feed water flow can cause serious impingement damage on pump impellors/veins etc.

3 BOILER CORROSION

As implied in the previous section the prime underlying villain that causes the serious corrosion problems in the boiler is dissolved oxygen. The complete nullification of dissolved oxygen, by physical and/or chemical means together with the maintenance concentrations of other necessary boiler water conditioning chemicals will ensure freedom from serious metal wastage. Metal deformation due to metal overheating, failure of boiler safety alarms etc. are secondary problems not covered here.

Cathodic reactions:

$$O_2 + 2H_2O + 4e \longrightarrow 4OH^-$$

$$2H^+ + 2e \longrightarrow H_2\uparrow$$

These are the two major reactions of concern. The presence of oxygen, apart from its cathodic reduction action, can also oxidise ferrous ions generated at the anode to the ferric state which precipitates as a hydrated oxide, $Fe_2O_3 \times H_2O$. This oxide does not offer the degree of metal protection required. By the total elimination of dissolved oxygen the cathodic electrochemical reactions are limited primarily to hydrogen ion reduction. The overall metal surface reaction under such water conditions can be recorded as:

$$3Fe + 4H_2O \longrightarrow Fe_3O_4 + 4H_2\uparrow$$

see Figure 3, Pourbaix diagram[1]

3.1 Oxygen Scavengers

The chemical reaction kinetics of the commercially employed oxygen scavengers are favoured by alkaline conditions.

The two most common additives employed as oxygen scavengers in the low boiler pressure market are sodium sulphite and tannins.

3.1.1 Sodium Sulphite. Finds application primarily in the lower pressure boiler plants (250 p.s.i.g.) but it can be used up to 600 p.s.i.g. (254°C). At this higher boiler pressure an awareness of the potential thermal decomposition should be noted.

The oxygen scavenging reaction is represented as:

$$2Na_2SO_3 + O_2 \longrightarrow 2Na_2SO_4$$

The reaction kinetics can be quite slow especially at colder water temperatures. The situation can be dramatically improved by the introduction of a catalyst, usually a transition metal salt; the salt most commonly employed is cobalt sulphate.

Sodium sulphite - catalyst reactions:

$$O_2 + 6CoSO_4 + 2H_2O \longrightarrow 2Co_2(SO_4)_3 + 2Co(OH)_2$$

$$Co_2(SO_4)_3 + Co(OH)_2 + Na_2SO_3 \longrightarrow Na_2SO_4 + 3CoSO_4 + H_2O$$

Catalysed sodium sulphite will reduce the dissolved oxygen content of a boiler feed water usually within 20 seconds. It is beneficial to remove all the dissolved oxygen from the feedwater before it enters the boiler. With the oxygen scavenging performance of catalysed sulphite this can be achieved. It is also a useful product where an economiser constitutes part of the feed system.

Apart from the oxygen reaction sodium sulphite, like most of the oxygen scavengers, will reduce $Fe_2O_3 \longrightarrow Fe_3O_4$.

On the negative side the thermal decomposition of sodium sulphite produces corrosive products, which can cause serious problems with turbine blades.

The net results of sodium sulphite decomposition are:

$$Na_2SO_3 + H_2O \xrightarrow[\text{heat}]{} 2NaOH + SO_2\uparrow$$

$$4Na_2SO_3 + 2H_2O \xrightarrow[\text{heat}]{} 3Na_2SO_4 + 2NaOH + H_2S$$

3.1.2 Tannins. These complex natural organics continue to receive wide acceptance in boiler water treatment. The tannins offer several useful properties in the overall boiler water conditioning programme. In this instance only the tannin's reducing properties are for consideration.

The products of tannin hydrolysis yield oxidisable polyhydroxyphenols, e.g. pyrogallol, gallic acid. These products undergo oxidation by the following route:

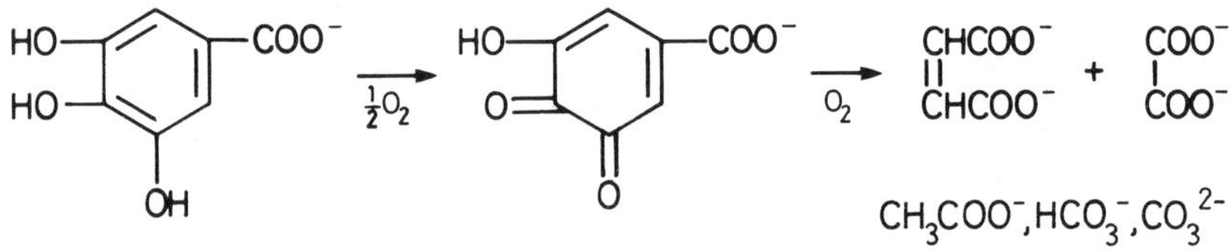

Gallic acid

Apart from their oxygen scavenging properties tannins also exhibit a dual corrosion inhibiting property. Firstly, they can beneficially affect the formation of a protective 'magnetite' iron oxide film and secondly selected hydrolysed residues can form 'iron tannates'.

The 'TANNINS' have an economical advantage over sodium sulphite in that they only contribute minimally to the boiler total dissolved solids, TDS. (TDS is a measurement used to control the degree of blowdown given from a boiler.)

3.1.3 Hydrazine. Hydrazine has been a most popular and accepted very affective oxygen scavenger/metal passivator. The product is inorganic and does not contribute to boiler TDS. These properties made hydrazine a most favoured product for use in the higher pressure boilers and still finds acceptance with the C.E.G.B.

The stigma of hydrazine having carcinogenic properties has seen a decline in the use of hydrazine as a boiler water additive. Where it is still used, or to be recommended, calls for the installation of suitable dosing equipment.

The oxygen scavenging performance of hydrazine can be greatly improved by the addition of hydroquinone as a catalyst.

Reactions:

$$N_2H_4 + O_2 \longrightarrow N_2 + 2H_2O$$

Hydrazine itself is not a particularly efficient oxygen scavenger (especially in cold conditions 25°C) and thus a surface reaction mechanism better describes its behaviour.

$$6Fe_2O_3 + N_2H_4 \longrightarrow 4Fe_3O_4 + N_2\uparrow + 2H_2O$$

$$4Fe_3O_4 + O_2 \longrightarrow 6Fe_2O_3$$

Hydroquinone catalysed:

$$2\ \text{Hydroquinone} + O_2 \longrightarrow 2\ \text{Quinone} + 2H_2O$$

$$2\ \text{Quinone} + N_2H_4 \longrightarrow 2\ \text{Hydroquinone} + N_2\uparrow$$

Hydrazine decomposition at higher boiler pressures can occur in accordance with the reaction:

$$2N_2H_4 \longrightarrow H_2\uparrow + N_2\uparrow + 2NH_3\uparrow$$

This breakdown of hydrazine offers no detriment to the boiler itself but the liberated ammonia in the presence of oxygen in the condensate can result in the corrosion of copper and copper alloys.

Included in this section of hydrazine one can add the commercially used chemical semicarbazide: $NH_2NHCONHNH_2$. This produces hydrazine upon hydrolysis.

3.1.4 Hydroquinone. The hydroquinone, quinone redox system differentiates hydroquinone from the polyhydroxyphenols produced from hydrolysed tannins, which only undergo the oxidation stages and are not reversible.

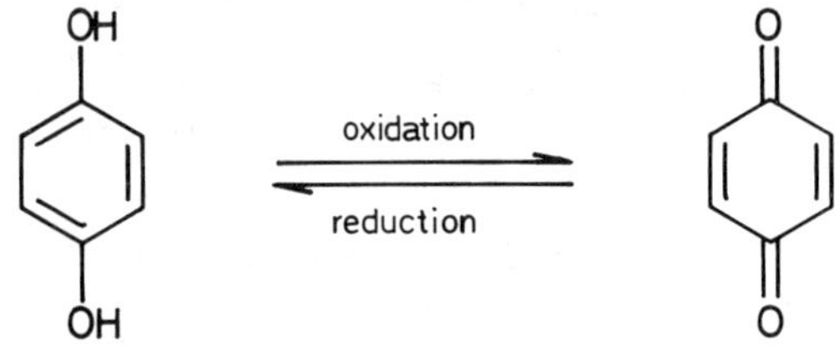

Hydroquinone also reacts with oxidized forms of iron and copper, which can be most useful when considering condensate line protection.

$$6Fe(OH)_3 + 2C_6H_4(OH)_2 — 2Fe_3O_4 + 2C_6H_4O_2 + 10H_2O$$

Hydroquinone is not steam volatile and thus to benefit from the products metal passivating properties in the after boiler section it has to be injected into the condensate system. For condensate applications and the preparation of commerically concentrated liquids, hydroquinone is usually blended with neutralising amines. A typical formulation would have a pH of 10.3.

In respect of product decomposition at the higher boiler pressure there is evidence that the products formed are light alcohols and ketones. There appears to be no problems arising from this hydrolysis.

3.1.5 Erythorbic Acid (iso ascorbic acid)[2]. Erythorbic acid is a strong reducing agent. Exposed aqueous solutions readily react with atmospheric oxygen.

For water treatment applications erythorbic acid is often supplied as the sodium or ammonium salt. Practical results have show the product to produce a good protective magnetite film on ferrous metals. The magnetite film formed in the presence of erythorbate is produced more quickly and offers better protection than a magnetite film produced in the presence of sulphite. There are several possible reactions that can occur between erythorbate and oxygen, but for boiler water calculation a stoichiometric ratio of 1:1 based on equivalent weights is used, since this will account for the fastest reaction.

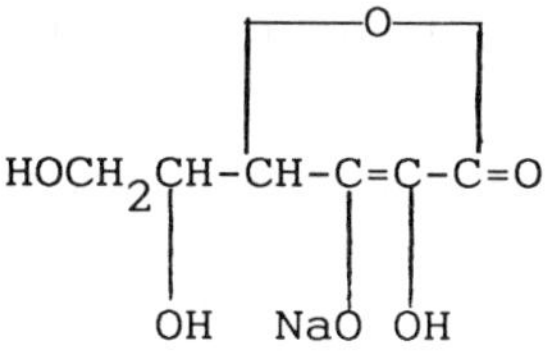

Sodium erythorbate
$NaC_6H_7O_6$

$$2(\text{erythorbate}) + O_2 \longrightarrow 2(\text{dehydroerythorbate}) + 2OH^-$$

Erythorbic acid is one of the few boiler water additives that has FDA approval status.

3.2 Volatile Oxygen Scavengers (VOS)

There are now products available which exhibit steam volatility and thus offer benefits as condensate line metal passivators. At the present time there are two products that fall into this category.

3.2.1 Diethylhydroxylamine (DEHA)[3]. Apart from its toxicological advantages versus hydrazine, DEHA offers a property of volatility from the boiler water into the condensate system.

The ability to scavenge oxygen and hence prevent its role in corrosion promotion in the boiler feed system, does not automatically negate oxygen corrosion in the condensate return lines. A product, such as DEHA, which will also be released from the boiler water into the condensate system will have the secondary benefits as a metal passivator and protector of the condensate system.

Thermal breakdown does not occur up to 300 p.s.i. Decomposition has been observed at 1000 p.s.i. with the formation of diethylamine and acetaldehyde. A probable mechanism is:

$$3(C_2H_5)_2NOH \longrightarrow 2(C_2H_5)NH + CH_3CH{=}NOH + CH_3CHO + H_2O$$

The reported DEHA oxidation products are primarily acetate and acetaldehyde in the lower pressure package boilers, whereas in the high pressure units diethylamine and acetaldehyde are predominant. Proposed reaction routes:

$$4(C_2H_5)_2NOH + 9O_2 \longrightarrow 8CH_3COOH + 2N_2 + 6H_2O$$

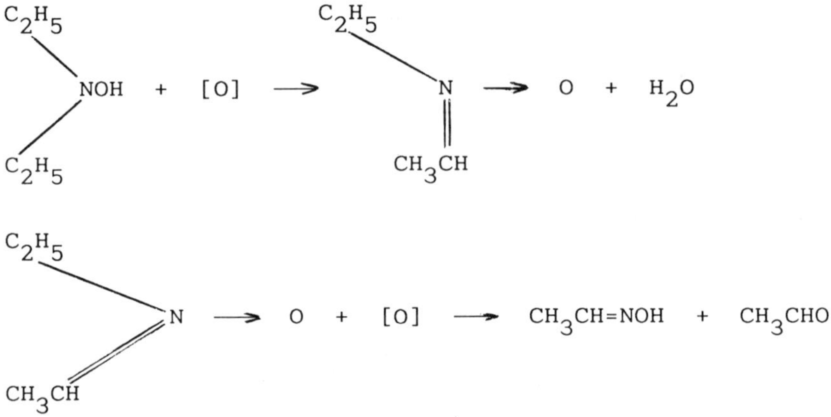

$$CH_3CH{=}NOH + O_2 \longrightarrow CH_3CHO + HNO_2$$

In package type boiler plant operating at low/medium pressure, there is advantage in mixing the VOS with catalysed sulphite. This combination uses the rapid oxygen scavenging properties of the catalysed sulphite, which would gain benefits in the protection of the boiler feed system, and release the DEHA to accelerate magnetite film formation on ferrous surfaces. The VOS property of DEHA would offer secondary benefits of condensate line protection without the need for further dosing points in the after boiler section.

Haematite reduction to magnetite is described by the equation:

$$(C_2H_5)_2N{=}OH + 6Fe_3O_4 \longrightarrow 4Fe_3O_4 + CH_3CH{=}NOH \quad CH_3CHOH \quad H_2O$$

<u>3.2.2 Methylethylketoxime (MEKO)</u>.[4,5] This is the latest VOS product to be introduced into the boiler water treatment market. It exhibits a higher distribution ratio than DEHA (see Figure 3.2.2 (a)) and is the least toxic of hydrazine organo 'alternatives' discussed (see Figure 3.2.2 (b)).

The introduction of MEKO has shown the product to behave in a similar manner to the other organo compounds in its ability to produce adherent protective magnetite films in all sections of the plant where ferrous metal constitutes the fabrication.

Oxygen scavenging is achieved in accordance with the following reaction:

$$CH_3\underset{\underset{NOH}{\|}}{C}C_2H_5 + O_2 \longrightarrow CH_3\underset{\underset{O}{\|}}{C}C_2H_5 + N_2O + H_2O$$

Metal passivation is described as:

$$2CH_3\text{-}\underset{\underset{NOH}{\|}}{C}\text{-}C_2H_5 + 6Fe_2O_3 \longrightarrow 4Fe_3O_4 + 2CH_3COC_2H_5 + N_2O + H_2O$$

4 CONDENSATE LINE CORROSION

In the condensate return line system both oxygen and carbon dioxide play important roles. The presence of dissolved oxygen usually manifests itself as pitting corrosion whereas carbon dioxide (carbonic acid) is identified as a very defined metal thinning.

The benefits of additives, as oxygen scavengers and metal passivators, in the condensate system have been reviewed. However, the performance of these products can be nullified if account is not taken of the acidic environment that carbonic acid can create.

Carbon dioxide/carbonic acid is primarily derived from the breakdown of bicarbonate alkalinity in the boiler feed water and also from the hydrolysis of carbonate, e.g. sodium carbonate.

$$2HCO_3^- \longrightarrow CO_3^{=} + CO_2 + H_2O$$

$$CO_3 + H_2O \longrightarrow 2OH^- + CO_2$$

Corrosion due to carbonic acid is described by the following equations:

$$CO_2 + H_2O \rightleftharpoons H_2CO_3 \rightleftharpoons H^+ + HCO_3^-$$

$$Fe + 2H^+ + 2HCO_3^- \longrightarrow Fe(HCO_3)_2 + H_2$$

In the presence of oxygen, ferrous oxidation can take place.

$$4Fe(HCO_3)_2 + O_2 \longrightarrow 2Fe_2O_3 + 8CO_2 + 4H_2O$$

The ferric oxide that precipitates can cause serious deposit and blockage problems in the condensate system. It should also be noted that carbon dioxide is again produced.

Condensate line problems can be experienced in all sections of the system from the first point of the steam phase change, e.g. turbine, to the far corners of a factory. Material failure can occur in process heating coils, overhead air heaters and steam calorifier with initial observed failures being noted at threaded joints and elbows. Apart from the financial losses incurred for material and equipment replacement, there is also a serious loss of condensate return to the boiler feed system, which, can also be an expensive experience. Another aspect of concern relates to health and safety. This can be simply exemplified by considering a large engineering workshop where the working conditions are kept satisfactory during the winter periods by the elevated air heaters. These units are usually copper tube heat exchangers having externally fitted fins. The heat energy is supplied conveniently by steam from the boiler. In the case of poor chemical conditioning, acidic conditions prevail and very quickly tube perforation occurs. The dropping of very hot water droplets onto unsuspecting workers below is not received with any dignity. The management are soon made aware of the situation and corrective actions quickly follow to rectify the situation and restore a working harmony.

4.1 the Role of Amines to Combat Acid Attack[6]

There are two types of 'amines' that find application in condensate line protection:

- NEUTRALISING
- FILMING

These names aptly describe the mechanism by which the two types of amines work.

4.1.1 Neutralising Amines. In the former class there are several products in current use. The selection of a particular product or product blend is usually made based on 'DISTRIBUTION RATIO' data (see Figure 3.2.2 (a)). This ratio describes the chemical presence in the vapour phase to that in the liquid phase. The amines showing a higher numerical value will give good condensate line protection at places more distant from the boiler than the amines having a lower value.

Morpholine would be a wise choice for protection of turbine blades and early sections of condensate lines whereas cyclohexylamine would be a better choice for the protection of distant pipework etc. It is a common practice to blend amines, with the two mentioned above being frequently used.

When selecting a neutralising amine programme where a deaerator is fitted, it should be remembered that the amines having the higher 'distribution ratio' will also give rise to higher losses at the deaerator.

Ammonia is a cheap strong base and neutralises carbonic acid. It can be added to the boiler feed system as an ammonium salt. In the alkaline boiler water conditions ammonia will be liberated which carries over into the condensate.

Ammonia may at first glance be a good choice product, but an awareness of the chemical's aggressive nature towards copper especially in the presence of oxygen should also be appreciated. The attack is via the monodentate ligand property of ammonia to form a soluble cupprammonium complex.

Amine dosage requirements are based purely on the carbon dioxide content and calls for one equivalent weight to neutralising amine per equivalent of carbonic acid.

e.g. Morpholine

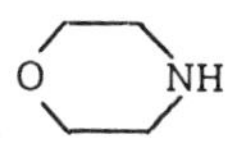

OC_4H_9N

$$OC_4H_9N + H_2CO_3 \longrightarrow OC_4H_9\overset{+}{N}H + HCO_3^-$$

Thus 1.98 parts of morpholine would be required to neutralise 1.0 part of carbonic acid, as CO_2. A selection of the neutralising amines are given in Figure 4.1. Food contact by live steam will very much control/limit the use of amines. Carbonic acid corrosion in this type of situation, or similar e.g. hospitals, can be greatly reduced by better control of the boiler feed water quality in terms of its bicarbonate content. This requirement can be achieved by the installation of a 'DE-ALKALISATION UNIT'. This simply comprises a weak ion exchange resin which converts any bicarbonate in the water into carbonic acid. A degasification step whilst the water is still slightly acidic releases CO_2.

4.1.2 Filming Amines. Of the filming amines the one that has generated most interest and application is OCTADECYLAMINE.

The filming amines, such as octadecylamine are prepared as 2.5-10% emulsions, as an acetate salt or slightly ethoxylated, e.g. ethoxylated soya amine, to aide the preparation of chemical solutions.

The variety of filming amines used as corrosion inhibitors in condensate return lines all exhibit a similar mechanistic behaviour in their corrosion protection performance. This similarity lies in the structure of the filming amines. There may be some slight variations in molecule structure but basically the amines are composed of a HYDROPHILIC and a HYDROPHOBIC component.

The hydrophilic element of the molecule relates to the water-liking (attracting) part of the structure, notably the amine head. The hydrophobic or water-hating (repelling) appertains to the aliphatic alkyl chain. The hydrophilic character can be changed by making an amine salt or introducing some ethoxylation, whilst the hydrophobic nature is adjusted by the control of the alkyl chain length.

Amine bonding to the metal surface is afforded by the donation of the lone pair of electrons on the nitrogen atom to the metal surface. Once the molecule has become adsorbed on the metal surface the molecule offers total hydrophobicity to the aqueous phase. The alkyl chain protrudes towards the aqueous phase repelling water from the metal surface. Amine adsorption can be readily identified since an exposed test piece when sprayed with water will produce water droplets on the surface.

In general 1-3 mg/l^{-1} of active amine is needed to give the overall protection required against both dissolved oxygen and carbon dioxide attack.

The application of filming amines may appear to be the answer to condensate return line protection; there are however certain aspects to be aware of.

The underdosing of a filming amine will give rise to areas that are not protected. These exposed areas will undergo rapid attack. The overdosing of a filming amine can give troubles with the build up of fatty amine deposits in the condensate system.

A further point to be noted is that the surface detergency action of the amine will result in the undermining of any corrosion debris causing disposal into other parts of the system where they will accumulate.

Filming/Neutralising Amine Combinations. There are many combinations of filming and neutralising amines that can be used. One of the more common mixtures being morpholine cyclohexylamine with octadecylamine.

5 SELECTED ORGANIC OXYGEN SCAVENGER WITH NEUTRALISING AMINE

This is becoming a popular approach to resolving the corrosion problems in condensate return line systems.

The combination of these types of products will obviously nullify the corrosive nature of the dissolved gases, oxygen and carbon dioxide. The reduced corrosion activity in the condensate system, apart from points raised earlier of condensate losses and health and safety, will reflect in reduced quantities of metal oxide and soluble metallic ions returning to the feed system. Further, the improved quality of the condensate may be sufficient to avoid condensate 'dumping' and also noticeably reduce the degree of ion exchange resin fouling if condensate polishing is practised.

From a physical chemistry aspect the addition of an alkaline neutralising amine could have supportive commercial reasons by increasing the solution strength of the oxygen scavenger/metal passivator, e.g. hydroquinone.

There is no basic limit to the number of products that can be used in combinations. Since the neutralising amines exhibit differences in their 'distribution ratio' it is possible (and practised) to blend 5 products together with the oxygen control additive.

6 HYDROXIDE ION/Ph CONTROL

The attainment of correct pH conditions in the various sections of the boiler plant have been partly discussed in respect of the condensate system. The introduction of alkaline chemicals for pH control serve several purposes:

- better oxygen scavenging results under alkaline conditions
- precipitation reactions in scale prevention
- corrosion reduction by hydroxide ion

The hydroxide ion can be both beneficial as a corrosion inhibitor, by aiding in the formation of protective oxide films on mild steel, and also corrosive. In this latter situation the hydroxide ion concentrates in localised parts of the boiler. In the past, caustic cracking (embrittlement) was more frequently encountered than it is in modern day boilers, but dissolution reactions can occur at cracks in oxide films and porous deposits where the hydroxide ion can concentrate; often termed 'ductile gouging'.

$$4NaOH + Fe_3O_4 \longrightarrow 2NaFeO_2 + Na_2FeO_2 + H_2O$$

$$Fe + NaOH + H_2O \longrightarrow NaFeO_2 + 3H \text{ (atomic hydrogen)}$$

The exposed metal tube would then react with boiler water to reform the magnetite film.

7 CHELANTS

Boiler water treatment additives themselves can be corrosive. The choice of a chelant as a complete antiscaling chemical or part of a programme has to be made with due care.

The main chelant used is EDTA with the product being dosed in stoichiometric quantities (or slightly less depending upon the selected programme) in respect of the divalent/trivalent ion concentrations.

Chelant attack can arise in:

- chelant feed line
- boiler feed pump, if chelant dosed on pressure side
- internal feed pipe and general boiler etching

Efficient oxygen scavenging, prior to chelant additions, is an essential requirement for this selected treatment. A regular analytical monitoring service has to be implemented when adopting a chelant programme.

8 SUMMARY

The water-side aspects of corrosion activity can be greatly reduced by:

- design
- product selection
- product dosing
- monitoring
- good working relationship with the chemical 'SERVICE COMPANY'

REFERENCES

1. V. Ashworth & P.J. Boden, Corros.Sci., 1970, 10, 709.

2. Pfizer, Chem.Div.Tech., Data Sheet 691.

3. D.M. Ellis, D.G. Cuisia, H.W. Thompson, 'The oxidation and degradation of volatile oxygen scavengers and their relevance in plant applications', presented at Corrosion/87, San Francisco, California, (Paper 432).

4. C.J. Capppabianca, P.F. Pelosi, J.H. Nulty, 'Experience with a Novel, Volatile Boiler Water Oxygen Scavenger', 46th Annual Meeting International Water Conference, Pittsburgh, Pennsylvania, November 1985.

5. P.F. Pelosi, J.S. Beecher, 'The Chemistry of Hydrazine and Methyl Ethyl Ketoxime under Hydrothermal Conditions', Presented at Corrosion/86, Houston, Texas.

6. J. Moulds, J.S. Beecher, 'Corrosion Control in Steam Generating Systems - The role of Sulfite, Hydrazine, Neutralising Amines and Ammonia', Water Chemistry and Corrosion Problems, PECS, Hungary, August 1981.

7. I.J. Cotton, Power, October 1980.

Corrosion Control by Surface Coatings; the Use of Chemical Inhibitors

By B. P. F. Goldie and D. G. Othen

BP RESEARCH CENTRE, SUNBURY-ON-THAMES, MIDDLESEX TW16 7LN, UK

1 INTRODUCTION

Metallic corrosion has an ongoing economic impact on everyday life and has been estimated to account for ~ 4% of GNP (approximately £2B annually in the UK) (1). Corrosion is due to a metal's inherent desire to return to a more chemically stable state when exposed to the environment. Metals vary in reactivity from those which for most conditions may be said to be inert (eg Au) to those which are highly reactive (eg. Na, K). Metals of engineering importance fall somewhere in between although in their pure form are more towards the reactive end of the range. However, engineering needs may not place corrosion as the first consideration in materials selection, more often it is the desired mechanical and structural properties which have to be primarily satisfied with the consequence that corrosion prevention measures have to be taken. Before considering the methods of corrosion prevention and control, it is necessary to look briefly at the process of corrosion itself.

2 BASIC CORROSION CONCEPTS

The majority of corrosion problems result from contact of a metal or alloy with a wet environment (ie in the presence of water). Pure water alone is for the most part non-corrosive particularly at ambient temperatures and pressures. It is the dissolved species such as ions (eg Cl^-, $SO_4^=$) and gases (eg O_2, CO_2) which determine the actual corrosivity of the wet environment. Such an environment is ionically conductive and a metal in contact with such an electrolyte will immediately develop a potential drop across the metal/solution interface. In an ideal situation, an equilibrium is set-up:-

$$M \rightleftharpoons M^{n+} + ne^- \qquad (1)$$

but the presence of other dissolved species influences this reaction, moving it out of equilibrium. Corrosion of metal occurs when the equilibrium is drawn to the right representing a steady dissolution of metal and the formation of metal ions (M^{n+}) in solution and electrons (ne^-) remaining in the bulk metal. On its own this represents a net charge imbalance within the metal which cannot exist in isolation, so for metal dissolution to occur the electrons produced must be consumed to maintain electrical neutrality. This is achieved by the presence of a second reaction at the metal/solution interface involving reduction of the dissolved species. These two reactions occur simultaneously forming the corrosion cell.

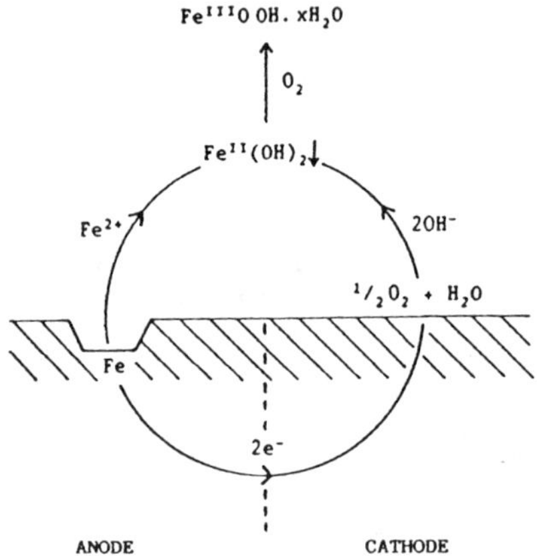

Figure 1 - Environmental Corrosion of Ferrous Metals

The reaction occurring at the cathode will depend on the pH of the electrolyte and the level of dissolved gases (oxygen). For example, the electrolytic corrosion of iron in an oxygenated neutral aqueous solution (Figure 1) can be written as:-

$$Fe \longrightarrow Fe^{2+} + 2e^- \quad (2)$$

$$^1/_2\ O_2 + H_2O + 2e^- \longrightarrow 2OH^- \quad (3)$$

$$Fe^{2+} + 2\ OH^- \longrightarrow Fe^{II}(OH)_2 \quad (4)$$

with subsequent air oxidation of the corrosion product to rust, variously written as $Fe^{III}OOH.xH_2O$ or $Fe_2{}^{III}O_3.nH_2O$.

It can be seen that, for corrosion to proceed, three steps are essential:- anodic oxidation (Equation 2), cathodic reduction (Equation 3), and ionic transfer through the electrolyte. Therefore corrosion can be prevented if any of these three steps is suppressed. This can be achieved in a number of ways, of which the use of surface coatings, and paints in particular, is the most common.

3 CORROSION PROTECTION BY PAINTS

Paints initially insulate a metal substrate from the water and oxygen required for environmental corrosion. However all organic film-forming polymers are to a greater or lesser extent permeable, and once saturation of the film has occurred their protective action can be envisaged in terms of three effects:-

(i) Resistive inhibition, ie placing a high resistance in the path of the corrosion current-carrying ions.

(ii) Anodic inhibition, ie preventing metal ions from leaving the substrate surface. A special case of this is the provision of a more reactive sacrificial anode, rendering the substrate entirely cathodic.

(iii) Cathodic inhibition, ie preventing access of oxygen to the cathode reaction sites.

In practical terms, this translates into three coating approaches (often combined in multi-coat systems):-

(i) Barrier coatings of low ionic permeability.

(ii) Zinc-rich primers, in which a very high loading of zinc dust permits sacrificial or galvanic protection, enhanced on ferrous substrates by the inhibitive properties of the zinc corrosion products.

(iii) Use of chemical inhibitors ("anti-corrosion pigments").

The addition of corrosion inhibitors to the paint film is probably the most common of all methods. These inhibitors, or anti-corrosion pigments, are commonly used in the form of solid particulate materials dispersed throughout the paint film. In this case, either (or both) of the cathodic or anodic reactions are suppressed for as long as inhibitor is present.

4 MECHANISMS OF CORROSION INHIBITORS

The fundamental mechanisms by which corrosion inhibitors in coatings operate are essentially those recognised for other applications, with the important proviso that the identity of the inhibiting species may not be immediately apparent; by their nature, corrosion inhibitors are reactive materials and reaction with other constituents of the coating may therefore occur. The corrosion reaction may be suppressed by one or more of the following processes:-

(i) pH control, ie provision of an alkaline environment to passivate the substrate.

(ii) Anodic inhibition.

(iii) Cathodic inhibition.

Choice of inhibitor is limited by the environment in which it must function, ie an organic polymer-based film adhesively bonded to a metal substrate. In particular, the integrity of the system (ie. cohesion and adhesion of the film) must be maintained for as long as possible. The inhibitor must therefore be compatible with the other film constituents; for example, few resin systems will tolerate highly alkaline materials, and the adhesive bonds to the substrate also tend to be alkali-sensitive (a common cause of failure by cathodic delamination once corrosion is established).

The presence of water-soluble material in a paint film is always potentially destructive; environmental exposure leads to swelling and/or blistering of the film from osmotic processes. There is therefore a drive towards low-solubility inhibitors, enhanced by the need to provide an inhibitor reservoir with a lifetime commensurate with that required for the coating system as a whole.

The release characteristics of an inhibitive pigment are therefore as important as its inhibitive efficiency; high solubility will lead to rapid depletion and destruction of the film by osmotic processes, whereas a very slow release rate will generate levels of inhibitor at the substrate too low for effective inhibition. Three release processes can be identified for inhibitive pigments, as follows:-

(i) Dissolution:-

The main disadvantage of solubility as a release mechanism is that the rate of release must be high enough to cope with aggressive environments. The inhibitor is therefore continuously leached out of the film, which has a limited lifetime. The only pigments which use solubility as their main release mechanism are the chromates.

(ii) Reaction - Acid/Base or Hydrolysis:-

Highly insoluble pigments such as zinc phosphate rely on hydrolysis or reaction with acid contaminants to release the anti-corrosive species. This process is essentially fortuitous and likely to be slow, although hydroxide from the cathodic corrosion reaction may aid hydrolysis.

(iii) Ion-Exchange:-

All insoluble metal oxides, hydroxides, and salts of oxyacids undergo surface ion-exchange in aqueous media. However, in most cases the surface area of active pigments is too low for this to be a significant release mechanism in the paint film. A notable exception is Red Lead, Pb_3O_4, where the open crystal structure allows Pb^{2+} release by an ion-exchange process.

The ideal corrosion inhibitor would be released at a rate dependent on the corrosive properties of its surroundings, allowing efficient use of the limited amount of inhibitor in the protective paint film. Clearly, solubility control alone cannot meet this requirement; dissolution rates will be affected only by temperature and common ion effects. Release by hydrolysis or acid/base reactions offers some degree of response to the surroundings, at least to acid contamination, and the corrosion process itself will to some extent stimulate hydrolytic release of an inhibitor through cathodic generation of hydroxide ion. However this does not provide a response to the non-acidic corrosion promoters such as chloride and sulphate ions, which are probably more important overall.

The ability of an aqueous solution to promote the electrolytic corrosion of a metal depends upon pH, the presence of corrosion promoters (usually chloride or sulphate ion), and the total ionic concentration, since ions carry the corrosion current. The corrosion reaction itself generates local increases in ionic concentration. Ion-exchange, the rate and extent of which will depend upon the concentration of electrolyte present, would therefore be an ideal process for controlling inhibitor release from an anti-corrosion pigment.

With these constraints in mind, it can be seen how the inhibitors used conform to the required inhibiting processes:-

(i) pH Control:-

Highly-alkaline pigments are rarely used; however, the majority are basic in reaction, which both lowers solubility and provides an acid/base reactive release mechanism.

(ii) Anodic Inhibition:-

Anodic inhibitors used in coatings are oxidative (chromate, molybdate) or film-forming (phosphate, borate), with the constraint that it must be possible to form a compound of comparatively low solubility. Thus nitrites, for example, are not used.

(iii) Cathodic Inhibitors:-

Cathodic inhibitors used in coatings are usually Group II metal cations (Mg^{2+}, Ca^{2+}, Sr^{2+}, Ba^{2+}, Zn^{2+}, Pb^{2+}), most commonly zinc, taking advantage of the range of medium to low solubility basic salts formed by these elements.

Organic film-forming inhibitors, although widely used in other applications, are not commonly used in corrosion-inhibiting coatings. However, such species are often important in their protective mechanisms. Environmental degradation of binder resins, particularly those based on or modified by drying oils, in the presence of basic Group II metal compounds generates soluble metal soaps (eg azelates), where the organic anion is an effective anodic inhibitor in its own right. This is also an important route for release of the cationic inhibitor itself.

Up until the emergence of environmental concerns in the 1960's, the common corrosion inhibitors used in coatings were based on lead compounds or chromates. Since then lead-based pigments in particular have been increasingly displaced by a wide and growing range of new non-toxic materials; more recently, chromates have also come under increasing pressure on toxicity grounds.

The most important traditional and new corrosion-inhibiting pigments are summarised in Tables 1 and 2 and discussed below:-

Lead-Based Pigments

The main lead-based anti-corrosion pigments used in coatings are Red Lead, Pb_3O_4, Calcium Plumbate, Ca_2PbO_4, and (mainly in the USA) Basic Lead Silicochromate - a complex material made by precipitating basic lead chromate onto a silica core. Lead compounds are generally held to be most effective in drying oil-based media (eg the classic Red Lead/Linseed Oil primers for poorly-prepared or rusted steel) due to the formation of inhibitive lead soaps, particularly lead azelate; however, this has not prevented their wider application in less reactive resins, which in itself argues for additional mechanisms. Pb^{2+} is conceptually a cathodic inhibitor by precipitation of $Pb(OH)_2$ at cathodic sites; however there is evidence that Pb^{2+} is in fact reduced to metallic lead (2), providing a surface more conducive to the reduction of oxygen and enabling a sufficiently high current density to be maintained for the formation of a protective γ-Fe_2O_3 oxide layer at anodic sites.

TABLE 1

TRADITIONAL ANTI-CORROSIVE PIGMENTS

Lead Group	Chromate Group
Red Lead	Basic Zinc Potassium Chromate
Calcium Plumbate	Strontium Chromate
	Zinc Tetraoxychromate
Basic Lead Silicochromate	

Others
Zinc Phosphate
Barium Metaborate

TABLE 2

THE MAJOR NON-TOXIC ANTI-CORROSIVE PIGMENTS

Phosphate Group	Molybdate Group
Zinc Phosphate	Basic Zinc Molybdate
Basic Zinc Phosphate	Calcium Zinc Molybdate
Basic Zinc Aluminium Phosphate	Basic Zinc Phosphate-Molybdate
Zinc Silicate-Modified Aluminium Tripolyphosphate	**Borate Group**
Zinc Hydroxyphosphite	Barium Metaborate
Complex Silicates	**Ion-Exchange Pigments**
Calcium Borosilicate	Calcium-Exchanged Silica

However, solubilisation of the inhibitor is still necessary; in drying-oil based media this is provided by the formation of lead soaps. Otherwise it is believed to be at least partially an ion-exchange process, aided by the tube-like crystal structure of Pb_3O_4, where Fe^{2+} ions from the corrosion reaction exchange for Pb^{2+} (2).

Chromate-Based Pigments

The principal chromate-based pigments are basic zinc potassium chromate, $K_2CrO_4.3ZnCrO_4.Zn(OH)_2.2H_2O$ (zinc yellow or zinc chrome), strontium chromate, $SrCrO_4$, and zinc tetraoxychromate, $ZnCrO_4.nZn(OH)_2$ ($n \sim 3$). The latter is principally used in wash or etch primers based on polyvinylbutyral, where its chromate content is largely consumed in the cure mechanism of the resin. Other Group II metal chromates are also used to a lesser extent, where different solubility characteristics are required.

The inhibitive mechanism of these species is the normal chromate mechanism (oxidative formation of a protective mixed oxide film, with defects plugged by Cr_2O_3 (2,3)), probably modified by the presence of the Group II cationic cathodic inhibitor counter ion. Inhibitor release is principally solubility-controlled, with an element of reactive release. As a result, despite their excellent inhibitive efficiency, chromate-pigmented primers tend to have restricted lifetimes under natural weathering conditions, particularly on steel.

Chromates are, however, still the principal inhibitive pigments for light alloys, until very recently there being no effective replacement available, see below. They are also still common for steel substrates, particularly in very high performance (ie low permeability) binders, or where long-term environmental exposure is not intended.

Non-Toxic Materials

The longest-established low toxicity anti-corrosive pigments are zinc phosphate, $Zn_3(PO_4)_2.2H_2O$ (in the UK and Europe), and barium metaborate (in the USA). Since toxicity in paints became a significant market influence, an abundance of other new materials has appeared. Table 2 lists the most important, ie those that have achieved a degree of market acceptance. Table 3 provides additional examples to indicate the range of inhibitors that have been proposed. Generally speaking they combine a film-forming anionic anodic inhibitor with a Group II cationic cathodic inibitor, often as a basic salt. Variants on zinc phosphate are common, as basic or mixed salts or using condensed or even reduced phosphates. By analogy with chromates, molybdates have been used

TABLE 3

FURTHER EXAMPLES OF NON-TOXIC ANTI-CORROSIVE PIGMENTS

Phosphate Group	Borate Group
Basic Zinc Carbonate Phosphate	Zinc Borate
Calcium Zinc Phosphate glasses	Calcium Borate
Zinc Tripolyphosphate	
Strontium Tripolyphosphate	**Complex Silicates**
Chromium Phosphate	Calcium Phosphosilicate
	Calcium Barium Phosphosilicate
Organic Salts	Zinc Phosphosilicate
Zinc Organophosphonates	Calcium Strontium Phosphosilicate
Organonitrogen Zinc Salts	
Zinc Benzoate	**Basic Pigments**
	Calcium Silicate
	Strontium Silicate
	Zinc Ferrite
	Calcium Ferrite
	Calcium Stannate
	Strontium Stannate
	Zinc Stannate

here as in other corrosion inhibiting applications, although they are comparatively weak oxidants. Phosphate-molybdate synergism has also been exploited. The solubilities of these materials are all fairly low to very low, ie their inhibitor release mechanisms are largely reactive.

Of these new materials, none has achieved the kind of market acceptance that would indicate a completely effective replacement for lead compounds or chromates; performance is generally held to be of the same overall order as zinc phosphate (itself long-established, at least in Europe, as the non-toxic alternative to lead or chrome pigments where lower performance is acceptable). Good, even excellent, performance can often be demonstrated for these materials, but appears to be very formulation-specific.

This is not unexpected. A reactive release mechanism is to some extent fortuitous and may be slow; better results will therefore be obtained if (for example) the resin contains reactive species, or its diffusion characteristics are well-matched to the release characteristics of the pigment. Additionally, the majority of the non-toxic anti-corrosive pigments are based on film-forming anionic inhibitors which require a considerably higher concentration at the metal surface to be effective than strongly oxidising inhibitors such as chromate; this will be more difficult to achieve with a slow release mechanism.

We therefore have two situations:- the combination of a highly efficient inhibitor with comparatively rapid release (eg chromates) which gives high levels of protection but is less suitable for extended environmental exposure, or the combination of a less efficient inhibitor with a comparatively slow release mechanism (zinc phosphate and most of the newer non-toxic pigments), which is less suitable for very aggressive conditions. In either case what is needed is a more efficient release mechanism providing a degree of response to conditions; ion-exchange can provide this. As was stated earlier, Red Lead has an ion-exchange capacity but otherwise this is not generally a significant release mechanism. However, pigments specifically designed to operate by this mechanism (eg calcium-exchanged silica) have recently become available (4).

Ion-Exchange Pigments

Ion-exchange pigments are comparatively high surface area inorganic oxides loaded with ionic corrosion inhibitors by ion-exchange with the surface hydroxyl groups. The oxides are chosen for their acidic or basic properties to provide cation- or anion-exchangers; thus silica is used as a cation support and alumina for anions. Examples from the patent literature are calcium- or zinc-exchanged silica, and chromate, phosphate, or nitrite-exchanged alumina. One advantage of this approach is to allow the use of otherwise highly soluble inhibitors. The only commercially-available material is the non-toxic calcium-exchanged silica.

The ion-exchange release mechanism allows efficient use of a comparatively low loading of inhibitive species; thus a Chromate/Alumina pigmented primer under natural weathering conditions out-performs an otherwise similar zinc chromate-based primer containing about ten times as much chromate.

The inhibitive mechanism of the ion-exchanged silicas is in fact more complex than simple ion-exchange, involving the silica support as well as the cation to form an inorganic barrier at the paint/metal interface. This is a true barrier to the current carrying ions and once formed, the corrosion reaction stops. Long-term protection and equivalence to chromates on light alloys has been demonstrated (5).

5 FUTURE DEVELOPMENTS

Barrier properties of coatings will undoubtedly be enhanced by advances in resin technology; whether this will render inhibitive pigments or other 'active' corrosion inhibition unnecessary for coatings remains to be seen.

In the meantime, development of inhibitive pigments will continue, particularly aimed at effective non-toxic chromate replacements. Low-solubility oxidative inhibitors such as vanadates are showing promise in some applications, for example. However, why inorganic pigments at all? Organic inhibitive pigments (mainly zinc salts) have been proposed, and in some cases even made available, for some time. It would also be quite possible to produce organic-soluble inhibitors of limited aqueous solubility (eg long-chain quaternary ammonium phosphates) which could be dissolved directly in the binder resin. Why have such ideas not met with wider acceptance?

The answer is probably two-fold; on the one hand, the conservatism of the paint industry, and on the other, the real practical point that the properties of a paint film are most strongly affected by those of the continuous phase, ie the binder resin. Within limits, changes in the dispersed phase (ie pigments and extenders) are less important than changes in the resin system - such as might be caused by an additive which was wholly or partially soluble in it. There is therefore a strong argument that inhibitors of this type would need to be developed for specific resin systems - or indeed, developed with them as a single package.

Development of corrosion-inhibiting and protective coatings will also be strongly affected by the results of research into the weathering of paint films and the corrosion process under coatings. Corrosion is going to be a problem well into the future and the need for protection even greater, providing the impetus for expanding R&D interests in this area.

ACKNOWLEDGEMENTS

The authors would like to thank the British Petroleum Company for permission to publish this paper.

REFERENCES

1. C. Cabrillac et al, Metals and Materials, 1987, 9, 533.

2. M. Svoboda and J. Mleziva, Progr. Org. Coatings, 1984, 12, 251.

3. H. Leidheiser Jr., J. Coatings Tech, 1981, 53, 29.

4. B.P.F. Goldie, Polymers Paint Colour J., 1985, 5, 337.

5. Awaiting publication.

Inhibitors for Use in Lubrication Service

By A. J. Gustavsen

KING INDUSTRIES, INC., 15133 ISLEVIEW DRIVE, BALLWIN, MO 63011, USA

1 INTRODUCTION

The primary purpose of a lubricant is to prevent damage to moving surfaces which are in contact. A lubricant accomplishes its purpose by forming a friction-reducing film between the surfaces, cooling the surfaces, reacting with a metal surface to form a lubricating product, cleaning debris from the surfaces, preventing corrosion of metal surfaces, or a combination of some or all of the above. There are many kinds of lubricants and many corrosion inhibitors used in them. This discussion will address only the oils and greases used as lubricants for operating engines and the most popular corrosion inhibitors used in such lubricants: sulfonates.

2 SULFONATES

Metal salts of organic sulfonic acids (sulfonates) have been used as corrosion inhibitors for many years. Although the primary function of a sulfonate in a lubricant or functional fluid like an hydraulic oil may be to prevent corrosion, sulfonates can also act as emulsifiers, demulsifiers, detergents, solubilisers for other additives and carriers of acid-neutralising carbonates.

Because of their effectiveness as corrosion inhibitors, the other benefits they can provide, and the synergism they exhibit with other inhibitors, sulfonates are one of the most common ingredients found in lubricants and functional fluids. There are, however, some problems associated with the use of sulfonates. Extreme pressure (EP) additives and sulfonates compete for the metal surface so that the concentration of EP additive must be increased to compensate. This increases the cost of the formulation as does the addition of more anti-oxidant which is necessary to overcome the pro-oxidant effect of sulfonates. Problems such as these are well known to formulators and easily overcome. Not easily overcome, however, are problems of sludge formation and the deposition of particulates on moving parts which occur with formulations containing sulfonates in high temperature applications.

The temperatures encountered in operating aircraft turbine engines are too high to permit the inclusion of sulfonates in lubricant formulations. The most widely used lubricant for United States Air Force (USAF) aircraft turbines is formulated to meet the MIL-L-7808J[1] (NATO Code O-148) specification.

This lubricant contains no sulfonates and does not provide sufficient corrosion protection for engines on aircraft which are inactive for prolonged periods or new and reconditioned engines which are in transit or storage. For such situations, the 7808 fluid is drained and replaced with MIL-C-8188C[2], a preservative oil having a composition similar to the 7808 lubricant, but containing a sulfonate corrosion preventive. Products qualified under the MIL-C-8188C specification fail to qualify under the MIL-L-7808J specification because of excessive sludge formation, foaming and deposition.

3 PRELIMINARY INVESTIGATION

The USAF does not want to continue using two fluids in its gas turbine engines. The need is especially critical in the Air Launched Cruise Missile (ALCM). It is extremely unlikely that any potential enemy planning a pre-emptive attack would be sporting enough to give us time to change the oil in all our missiles before they started firing theirs. Under USAF sponsorship, a corrosion inhibited lubricant for ALCM engines was developed.[3] This author was an unofficial consultant on the formulation of the new lubricant and supplied the samples which were the key to the successful completion of the project.

The United States Navy (USN) uses only one oil in its gas turbine engines, MIL-L-23699C.[4] They have been trying to make a corrosion inhibited version and decided to try the inhibitor package that had been developed for the ALCM. The inhibitor package was a two component mixture which, in the USN formulation formed a precipitate during aging tests. The precipitate was a reaction product of the two components and was oil insoluble. We found that we could react the two components in a concentrated solution and obtain a clear product which was stable in all of the lubricant base stocks required. This product (Product A) passed all USN laboratory tests and is now in engine testing. The same product is also being used in the further qualifying tests of the ALCM lubricant by the USAF.

At the same time that we were working on the USN project we started a study of the effect of high temperatures on sulfonates. As a sulfonate manufacturer, we realised that lubricants were being used at ever higher temperatures and that, eventually, someone would develop a corrosion inhibitor that would replace the thermally unstable sulfonates in lubricants. In self defense, we decided to try to be the developer of such a new corrosion inhibitor. We began by looking at how and why sulfonates decompose thermally.

The synthetic sulfonates used in lubricants are made from alkaryl sulfonic acids where the alkaryl moiety is, typically, dinonylnaphthalene or didodecylbenzene. It had been suggested that sulfonate decomposition began with the alkyl side chains and that more stable sulfonates might be obtained by using different alkyl groups. To explore this possibility, we modified a micro viscometer apparatus by replacing the glass sample holder with a test tube having a ground glass joint into which we placed a condenser. The viscometer apparatus was designed to heat the glass sample holder, and now our test tube, and to hold a specified temperature to +/- $1^{\circ}C$. The test tube held about 20 ml of oil. Since we were using concentrated sulfonates (50 % active), we had a large enough sample to do complete analyses on the thermally stressed products. Stressing was done with no agitation or aeration and was normally 21 hrs at temperature.

We found that the alkyl side chain was not a significant factor in the loss of sulfonate due to thermal degradation. The Lewis acidity of the metal ion portion of the sulfonate was found to be very significant. At a particular temperature, zinc or lead sulfonates would show much more loss of sulfonate content than would calcium sulfonates, but barium or sodium sulfonates would show less sulfonate loss than the calcium. Desulfonation reactions are generally known to be acid catalysed and to involve a stoichiometric amount of water.[5] Our initial work was done with barium sulfonates at 200°C. We found that sulfonate loss was 20 - 25%. The moisture content of the sulfonates was 0.5% which should have been sufficient for a 30% loss of sulfonate, if one mole of water was consumed for each mole of sulfonate destroyed. Although we soon found that the correlation was mostly coincidental, we briefly entertained the hope that desulfonation was water limited. Unfortunately, the water inside sulfonate micelles is very tightly held. Normal manufacturing practices involve drying under vacuum at 150°C and the lowest water contents obtained are 0.2 - 0.6%. Attempts to further decrease the water content by distillation led to decomposition of the sulfonate.

As mentioned above, we solubilised the reaction product of the two components in Product A by running the reaction in a concentrated solution. This suggested that the reaction product formed inside the sulfonate micelles. One reason we suspected that this was the case was that the water content of the product was so low. The water inside sulfonate micelles can be reversibly displaced by other polar materials such as metal carbonates (as in overbased sulfonates). The water content of Product A was 0.04%.

Stressing of Product A at 200°C showed no loss of sulfonate. Similar tests on low-water-containing overbased products gave results that were nearly as good and certainly far better than the neutral sulfonates first tested. Next we added one mole of water per mole of sulfonate to a sample of an overbased sulfonate and retested it. The overbased product first formed a heavy precipitate (the metal carbonate was displaced from the micelles) and, after 21 hrs at 200°C, analysis showed that a significant portion of the sulfonate had been destroyed. When we added water to Product A, however, no precipitate formed. This indicated that the reaction product was not displaced from the micelles by water. After 21 hrs at 200°C, we detected no loss of sulfonate! Obviously, Product A deserved further investigation.

4 PRODUCT B

In all of the above stressing experiments, the sulfonates and Product A had been in synthetic vehicles. This was because the USAF and USN were working with synthetic fluids based on esters, polyalphaolefins (PAO) or polyhalogenated materials (CTFE). We wanted to be able to run comparisons of Product A with other sulfonates, but most commercial sulfonates are only available in a mineral oil base. We, therefore, prepared an equivalent to Product A in a solvent extracted coastal oil and it will be designated as Product B. Products A and B are identical except for the vehicle base oils.

Product B, like Product A, showed no loss of sulfonate after stressing at 200°C for 21 hrs. We extended the time to 60 hrs and there was still no sulfonate loss. We went back to the 21 hrs testing and started increasing the temperature. At 215,225,240 and 260°C, we found no loss of sulfonate. Finally, at 300°C, the limit of our apparatus, we found only 97.7% of the original sulfonate still active. We had destroyed 2.3% of it. One of the components of Product B is a barium dinonylnaphthalene sulfonate. When the barium sulfonate part alone was stressed at 200°C for 21 hrs, 53.8% of the sulfonate was destroyed. Another synthetic sulfonate (a barium didodecylbenzene sulfonate) showed a loss of 51% of its starting sulfonate content during 21 hr exposure to 200°C. A blend of four parts of the barium didodecylbenzene sulfonate and one part of Product B lost none of its sulfonate content in 21hrs at 200°C. Product B stablises other sulfonates!

In all of the above work, we had been analysing for sulfonates using ASTM D-3049. This is an excellent analytical procedure, but it does not tell whether the material analysed is a good corrosion inhibitor or not. The primary function of the sulfonate is to prevent corrosion, so the question arose as to how Product B performed after thermal stressing. We decided to run some ASTM B-117 salt spray tests on oil dilutions of Product B and the barium dinonylnaphthalene sulfonate. The oil solutions were prepared using 10% of the standard forms of the ingredients. This meant that the actual sulfonate content of the sulfonate-only solution was 5% and the sulfonate content of the Product B solution was 3.6% although the total active content was 5.7% (the non-oil content of Product B is 57%). Steel panels were dipped in the solutions and subjected to salt spray testing. The solutions were subjected to thermal stressing at 200°C for 21 hrs and panels were dipped in the stressed oil and also subjected to salt spray testing. The hours to failure are shown in Table 1. The salt spray results confirm that the loss of sulfonate gave a corresponding loss of corrosion protection in the case of the plain sulfonate and no loss of corrosion protection with the stressed Product B. We can also see that

Table 1 Hours of ASTM B-117 Salt Spray Before Failure

Oil Solution	Unstressed	Stressed
Sulfonate Alone	4	1
Product B	9	9

Product B protected the metal from salt corrosion more than twice as long as sulfonate alone in the unstressed tests. This is not that surprising since both ingredients in Product B are corrosion inhibitors which are, fortunately, synergistic.

The results obtained from Products A and B showed that we had developed a method of extending the operational range of corrosion inhibiting sulfonates to much higher temperatures and the method worked with different kinds of sulfonates and in different vehicles. However, this had only been shown for barium sulfonates. Other sulfonate salts are more sensitive to thermal degradation than barium sulfonates. It remained to be seen if the same method would be effective with non-barium sulfonates.

5 OTHER METAL ANALOGS OF PRODUCT B

The preparation of lithium, calcium, magnesium and zinc analogs of Product B was not as straightforward as we would have wished. It was necessary to develop procedures and even modify the non-sulfonate portion before all of the desired products were successfully prepared. This is understandable when these products are compared to overbased sulfonates where different procedures and promoters are used for each metal sulfonate. Suffice to say, the desired products have been prepared and they are all like Products A and B in thermal stability. Rather than going into tedious detail on all of the products prepared, one example will provide the gist.

As mentioned above, zinc sulfonates are more thermally unstable than calcium sulfonates which, in turn, are less stable than barium sulfonates. When thermally stressed at 175°C for 21 hrs, zinc dinonylnaphthalene sulfonate lost 88% of its sulfonate. The zinc product we prepared (Product Z) showed no loss of sulfonate when stressed at 200°C for 21 hrs and, when the stressed material was diluted in oil to a 10% concentration, gave the salt spray results shown in Table 2. Note that Product Z gives more than three times longer protection than the zinc sulfonate alone. The same is

TABLE 2 Hours of ASTM B-117 Salt Spray Before Failure Stressed as Concentrates, Tested at 10% in Oil

Oil Solution	Unstressed	Stressed
Zinc Sulfonate Alone	5	1
Product Z	16	16

found with the other metal sulfonates and with a new, modified version of Product B. This increased activity allows lubricant formulators to use less additive to achieve a desired level of corrosion protection.

These products have been made available as samples to lubricant formulators during the last few months and we have gotten some very interesting feedback. Most commonly used additives for lubricants are compatible with the new products. When used as replacements for sulfonates, significantly lower concentrations of the new products usually give equivalent corrosion protection, although, in a few formulations, the new products were only effective at the same concentration as the standard sulfonates.

The new products appear to act just like sulfonates when it comes to competing for the metal surface with EP additives, so the standard practice of adding a little more EP additive when a corrosion inhibitor is added is still valid. The situation is quite different, however, when it comes to anti-oxidants. Sulfonates, especially overbased sulfonates, are notorious pro-oxidants and formulators expect increases in sludge formation and acid values in oxidation tests when sulfonates are added to formulations. When the new products are added to lubricants, sludge formation actually decreases as does deposition. Some formulators report being able to reduce the amount of anti-oxidant they use and others report that they are able to use less expensive anti-oxidants when the new products are added to formulations. The pro-anti-oxidant effect was quite unexpected and we are currently trying to learn more about it. This author suspects that the effect is really a case of the new products stabilising the other additives (EP agents, VI improvers, etc.) in the lubricants.

6 CONCLUSION

A new series of sulfonate based corrosion inhibitors has been developed. These inhibitors are thermally stable and functional at temperatures where ordinary sulfonates decompose. The inhibitors also show a pro-anti-oxidant effect which is being further investigated. In some cases, the corrosion inhibition provided by these new products is quite significantly better than that provided by like concentrations of ordinary sulfonates.

REFERENCES

1. MIL-L-7808J, Military Specification - Lubricating Oil, Aircraft Turbine Engine, Synthetic Base, 11 May 1982.
2. MIL-C-8188C, Military Specification - Corrosion Preventive Oil, Gas Turbine Engine, Aircraft Synthetic Base, 26 May 1959.
3. P.A. Warner and W.E. Ward, Lubr. Eng., 1985, 41, 348.
4. MIL-L-23699C, Military Specification - Lubricating Oil, Aircraft Turbine Engine, Synthetic Base, 4 October 1978.
5. E.E. Gilbert, "Sulfonation and Related Reactions", Interscience, New York, 1965.